In Intellectual Property in the 21st Century

PATENTS, TECHNOLOGY AND COMMERCE

INTELLECTUAL PROPERTY IN THE 21ST CENTURY SERIES

Patents, Technology and Commerce
Christie L. Jansen and Derek A. Hyman (Editors)
2009. ISBN 978-1-60692-291-0

In Intellectual Property in the 21st Century

PATENTS, TECHNOLOGY AND COMMERCE

CHRISTIE L. JANSEN
AND
DEREK A. HYMAN
EDITORS

Nova Science Publishers, Inc.
New York

Copyright © 2009 by Nova Science Publishers, Inc.

All rights reserved. No part of this book may be reproduced, stored in a retrieval system or transmitted in any form or by any means: electronic, electrostatic, magnetic, tape, mechanical photocopying, recording or otherwise without the written permission of the Publisher.

For permission to use material from this book please contact us:
Telephone 631-231-7269; Fax 631-231-8175
Web Site: http://www.novapublishers.com

NOTICE TO THE READER
The Publisher has taken reasonable care in the preparation of this book, but makes no expressed or implied warranty of any kind and assumes no responsibility for any errors or omissions. No liability is assumed for incidental or consequential damages in connection with or arising out of information contained in this book. The Publisher shall not be liable for any special, consequential, or exemplary damages resulting, in whole or in part, from the readers' use of, or reliance upon, this material.

Independent verification should be sought for any data, advice or recommendations contained in this book. In addition, no responsibility is assumed by the publisher for any injury and/or damage to persons or property arising from any methods, products, instructions, ideas or otherwise contained in this publication.

This publication is designed to provide accurate and authoritative information with regard to the subject matter covered herein. It is sold with the clear understanding that the Publisher is not engaged in rendering legal or any other professional services. If legal or any other expert assistance is required, the services of a competent person should be sought. FROM A DECLARATION OF PARTICIPANTS JOINTLY ADOPTED BY A COMMITTEE OF THE AMERICAN BAR ASSOCIATION AND A COMMITTEE OF PUBLISHERS.

LIBRARY OF CONGRESS CATALOGING-IN-PUBLICATION DATA
Available upon request
ISBN: 978-1-60692-291-0

Published by Nova Science Publishers, Inc. ✛ New York

CONTENTS

Preface		vii
Chapter 1	Industrial Competitiveness and Technological Advancement: Debate over Government Policy *Wendy H. Schacht*	1
Chapter 2	Cooperative R&D: Federal Efforts to Promote Industrial Competitiveness *Wendy H. Schacht*	29
Chapter 3	The Bayh-Dole Act: Selected Issues in Patent Policy and the Commercialization of Technology *Wendy H. Schacht*	55
Index		89

PREFACE

There is an ongoing interest in the pace of U.S. technological advancement due to its influence on U.S. economic growth, productivity and international competitiveness. Because technology can contribute to economic growth and productivity increases, congressional attention has focused on how to augment private-sector technological development. This book addresses questions that have been raised concerning the proper role of the federal government in technology development and the competitiveness of U.S. industry. In response to the foreign challenge in the global marketplace, the United States Congress has explored ways to stimulate technological advancement in the private sector. Among the issues addressed in this book are whether joint ventures contribute to industrial competitiveness and what role, if any, the government has in facilitating such arrangements. This book also discusses the Bayh-Dole Act, which grants patent rights to inventions arising out of government-sponsored research and development to certain types of entities.

Chapter 1 - Legislative activity over the past two decades has created a policy for technology development, albeit an ad hoc one. Because of the lack of consensus on the scope and direction of a national policy, Congress has taken an incremental approach aimed at creating new mechanisms to facilitate technological advancement in particular areas and making changes and improvements as necessary.

Congressional action has mandated specific technology development programs and obligations in federal agencies that did not initially support such efforts. Many programs were created based upon what individual committees judged appropriate within the agencies over which they had authorization or appropriation responsibilities. The use of line item funding for these activities, including the Advanced Technology Program and the Manufacturing Extension Program of the National Institute of Standards and Technology, as well as for the Undersecretary for

Technology at the Department of Commerce, is viewed by proponents as a way to ensure that the government encourages technological advance in the private sector.

Some legislative activity, beginning in the 104th Congress, has been directed at eliminating or significantly curtailing many of these federal efforts. Although this approach has not been adopted, the budgets for several programs have declined. Questions have been raised concerning the proper role of the federal government in technology development and the competitiveness of U.S. industry. As the 110th Congress begins to develop its budget priorities, how the government encourages technological progress in the private sector again may be explored and/or redefined.

Chapter 2 - In response to the foreign challenge in the global marketplace, the United States Congress has explored ways to stimulate technological advancement in the private sector. The government has supported various efforts to promote cooperative research and development activities among industry, universities, and the federal R&D establishment designed to increase the competitiveness of American industry and to encourage the generation of new products, processes, and services.

Collaborative ventures are intended to accommodate the strengths and responsibilities of all sectors involved in innovation and technology development. Academia, industry, and government often have complementary functions. Joint projects allow for the sharing of costs, risks, facilities, and expertise.

Cooperative activity covers various institutional and legal arrangements including industry-industry, industry-university, and industry-government efforts. Proponents of joint ventures argue that they permit work to be done that is too expensive for one company to support and allow for R&D that crosses traditional boundaries of expertise and experience. Such arrangements make use of existing, and support the development of new, resources, facilities, knowledge, and skills. Opponents argue that these endeavors dampen competition necessary for innovation.

Federal efforts to encourage cooperative activities include the National Cooperative Research Act; the National Cooperative Production Act; tax changes permitting credits for industry payments to universities for R&D and deductions for contributions of equipment used in academic research; and amendments to the patent laws vesting title to inventions made under federal funding in universities. Technology transfer from the government to the private sector is facilitated by several laws. In addition, there are various ongoing cooperative programs supported by various federal departments and agencies.

Given the increased popularity of cooperative programs, questions might be raised as to whether they are meeting expectations. Among the issues before

Congress are whether joint ventures contribute to industrial competitiveness and what role, if any, the government has in facilitating such arrangements.

Chapter 3 - Congressional interest in facilitating U.S. technological innovation led to the passage of P.L. 96-5 17, Amendments to the Patent and Trademark Act (commonly referred to as the Bayh-Dole Act after its two main sponsors). The act grants patent rights to inventions arising out of government-sponsored research and development (R&D) to certain types of entities with the expressed purpose of encouraging the commercialization of new technologies through cooperative ventures between and among the research community, small business, and industry.

Patents provide an economic incentive for companies to pursue further development and commercialization. Studies indicate that research funding accounts for approximately one-quarter of the costs associated with bringing a new product to market. Patent ownership is seen as a way to encourage the additional, and often substantial investment necessary for generating new goods and services. In an academic setting, the possession of title to inventions is expected to provide motivation for the university to license the technology to the private sector for commercialization in expectation of royalty payments.

The Bayh-Dole Act has been seen as particularly successful in meeting its objectives. However, while the legislation provides a general framework to promote expanded utilization of the results of federally funded research and development, questions are being raised as to the adequacy of current arrangements. Most agree that closer cooperation among industry, government, and academia can augment funding sources (both in the private and public sectors), increase technology transfer, stimulate more innovation (beyond invention), lead to new products and processes, and expand markets. However, others point out that collaboration may provide an increased opportunity for conflict of interest, redirection of research, less openness in sharing of scientific discovery, and a greater emphasis on applied rather than basic research. Additional concerns have been expressed, particularly in relation to the pharmaceutical and biotechnology industries, that the government and the public are not receiving benefits commensurate with the federal contribution to the initial research and development.

Actual experience and cited studies suggest that companies which do not control the results of their investments — either through ownership of patent title, exclusive license, or pricing decisions — tend to be less likely to engage in related R&D. The importance of control over intellectual property is reinforced by the positive effect P.L. 96-517 has had on the emergence of new technologies and techniques generated by U.S. companies.

Chapter 1

INDUSTRIAL COMPETITIVENESS AND TECHNOLOGICAL ADVANCEMENT: DEBATE OVER GOVERNMENT POLICY[*]

Wendy H. Schacht

ABSTRACT

There is ongoing interest in the pace of U.S. technological advancement due to its influence on U.S. economic growth, productivity, and international competitiveness. Because technology can contribute to economic growth and productivity increases, congressional attention has focused on how to augment private-sector technological development. Legislative activity over the past two decades has created a policy for technology development, albeit an ad hoc one. Because of the lack of consensus on the scope and direction of a national policy, Congress has taken an incremental approach aimed at creating new mechanisms to facilitate technological advancement in particular areas and making changes and improvements as necessary.

Congressional action has mandated specific technology development programs and obligations in federal agencies that did not initially support such efforts. Many programs were created based upon what individual committees judged appropriate within the agencies over which they had authorization or appropriation responsibilities. The use of line item funding for these activities, including the Advanced Technology Program and the Manufacturing Extension Program of the National Institute of Standards and Technology, as well as

---

[*] Excerpted from CRS Report RL33528, dated August 1, 2007.

for the Undersecretary for Technology at the Department of Commerce, is viewed by proponents as a way to ensure that the government encourages technological advance in the private sector.

Some legislative activity, beginning in the 104th Congress, has been directed at eliminating or significantly curtailing many of these federal efforts. Although this approach has not been adopted, the budgets for several programs have declined. Questions have been raised concerning the proper role of the federal government in technology development and the competitiveness of U.S. industry. As the 110th Congress begins to develop its budget priorities, how the government encourages technological progress in the private sector again may be explored and/or redefined.

MOST RECENT DEVELOPMENTS

Congressional initiatives have often been relied on over the past 25 years to support technological advancement in U.S. industry. This approach has involved both direct measures that concern budget outlays and the provision of services by government agencies (such as the Advanced Technology Program (ATP) and the Manufacturing Extension Partnership (MEP) of the National Institute of Standards and Technology) and indirect measures that include financial incentives and legal changes. Many of these efforts, however, have been revisited since the 104th Congress given the then Republican majority's statements in favor of indirect strategies such as tax policies, intellectual property right protection, and antitrust laws to promote technological advancement; increased government support for basic research; and decreased direct federal funding for private sector technology initiatives. Beginning in FY2000, the original House-passed appropriation bills have not included funding for ATP. In addition, the President's FY2003 budget for the first time requested a significant reduction in support for MEP based on the idea that all manufacturing extension centers operating more than six years should continue without federal funding. While no program has been eliminated, several have been financed at reduced levels.

P.L. 110-5, enacted in the 110th Congress, provides FY2007 appropriations of $104.6 million for MEP and $79 million for ATP. The President's FY2008 budget proposes a significant decrease in support for manufacturing extension to $46.3 million and includes no funding for ATP. H.R. 3093, as passed by the House, would provide $108.8 million for MEP and $93.1 million for ATP. S. 1745, as reported from the Senate Committee on Appropriations, would finance MEP at $110 million and fund ATP at $100 million (with a $10 million rescission from these two activities). The Manufacturing Technology Competitiveness Act of 2007, H.R. 255, establishes several new manufacturing technology programs for small and medium- sized firms. Appropriations for MEP through FY20 12 would be authorized by S. 69. The Technology Innovation and Manufacturing Stimulation Act of 2007, H.R. 1868, as passed by the House, authorizes funding for NIST through FY20 10 and creates several new manufacturing R&D programs in that organization as does H.R. 2272, the 21st Century Competitiveness Act of 2007, as agreed to in the conference report, in addition to creating a new Technology Innovation Program to replace ATP (among other things).

Several of the actions detailed in the "American Competitiveness Initiative" announced by the President in the 2006 State of the Union Address are included in bills introduced in the 110th Congress. The ACI proposed various innovation-related activities including increased basic research funding, making permanent the research and experimentation tax credit (which was extended through the end of 2007 by P.L. 109-432), and improved math and science education. S. 833, the Competitiveness Through Education, Technology, and Enterprise Act of 2007, would make the research tax credit permanent, as does S. 41, the Research Competitiveness Act of 2007, and H.R. 1712, the Research and Development Tax Credit Act of 2007, which also create tax exempt facility bonds for the development of research park facilities, among other things. S. 592 extends the research credit through 2012. H.R. 85, the Energy Technology Transfer Act, as passed by the House, would establish a program of grants to non-profit institutions, state and local governments, cooperative extension services, or universities to transfer energy efficient methods and technologies. H.R. 363, the Sowing the Seeds Through Science and Engineering Research Act, as passed by the House, authorizes a Presidential Innovation Award, among other things. S. 761, the America Creating Opportunities to Meaningfully Promote Excellence in Technology, Education, and Science Act, passed by the Senate and incorporated as an amendment to H.R. 2772 (see above), would authorize appropriations for NIST through FY20 11, as well as providing for the creation of several programs, studies, and initiatives designed to improve U.S. innovation and competitiveness.

BACKGROUND AND ANALYSIS

Technology and Competitiveness

Interest in technology development and industrial innovation increased as concern mounted over the economic strength of the nation and over competition from abroad. For the United States to be competitive in the world economy, U.S.

companies must be able to engage in trade, retain market shares, and offer high quality products, processes, and services while the nation maintains economic growth and a high standard of living. Technological advancement is important because the commercialization of inventions provides economic benefits from the sale of new products or services; from new ways to provide a service; or from new processes that increase productivity and efficiency. It is widely accepted that technological progress is responsible for up to one-half the growth of the U.S. economy, and is one principal driving force in long-term growth and increases in living standards.

Technological advances can further economic growth because they contribute to the creation of new goods, new services, new jobs, and new capital. The application of technology can improve productivity and the quality of products. It can expand the range of services that can be offered as well as extend the geographic distribution of these services. The development and use of technology also plays a major role in determining patterns of international trade by affecting the comparative advantages of industrial sectors. Since technological progress is not necessarily determined by economic conditions — it also can be influenced by advances in science, the organization and management of firms, government activity, or serendipity — it can have effects on trade independent of shifts in macroeconomic factors. New technologies also can help compensate for possible disadvantages in the cost of capital and labor faced by firms.

Federal Role

In the recent past, American companies faced increased competitive pressures in the international marketplace from firms based in countries where governments actively promote commercial technological development and application. In the United States, the generation of technology for the commercial marketplace is primarily a private sector activity. The federal government traditionally becomes involved only for certain limited purposes. Typically these are activities which have been determined to be necessary for the "national good" but which cannot, or will not, be supported by industry.

To date, the U.S. government has funded research and development (R&D) to meet the mission requirements of the federal departments and agencies. It also finances efforts in areas where there is an identified need for research, primarily basic research, not being performed in the private sector. Federal support reflects a consensus that basic research is critical because it is the foundation for many new innovations. However, any returns created by this activity are generally long term, sometimes not marketable, and not always evident. Yet the rate of return to society as a whole generated by investments in research is significantly larger than the benefits that can be captured by the firm doing the work.[1]

Many past government activities to increase basic research were based on a "linear" model of innovation. This theory viewed technological advancement as a series of sequential steps starting with idea origination and moving through basic research, applied research, development, commercialization, and diffusion into the economy. Increases in federal funds in the basic research stage were expected to result in concomitant increases in new products and processes. However, this linear concept is no longer considered valid. Innovations often occur that do not require basic or applied research or development; in fact most innovations are incremental improvements to existing products or processes. In certain areas, such as biotechnology, the distinctions between basic research and commercialization are small and shrinking. In others, the differentiation between basic and applied research is artificial. The critical factor is the commercialization of the technology. Economic benefits accrue only when a technology or technique is brought to the marketplace where it can be sold to generate income or applied to increase productivity. Yet, while the United States has a strong basic research enterprise, foreign firms appear equally, if not more, adept at taking the results of these scientific efforts and making commercially viable products. Often U.S. companies are competing in the global marketplace against goods and services developed by foreign industries from research performed in the United States. Thus, there has been increased congressional interest in mechanisms to accelerate the development and commercialization processes in the private sector.

The development of a governmental effort to facilitate technological advance has been particularly difficult because of the absence of a consensus on the need for an articulated policy. Technology demonstration and commercialization have traditionally been considered private sector functions in the United States. While over the years there have been various programs and policies (such as tax credits, technology transfer to industry, and patents), the approach had been ad hoc and uncoordinated. Much of the program development was based upon what individual committees judged appropriate for the agencies over which they have jurisdiction. Despite the importance of technology to the economy, technology-related considerations often have not been integrated into economic decisions.

There have been attempts to provide a central focus for governmental activity in technology matters. P.L. 100-5 19 created within the Department of Commerce a Technology Administration headed by a new Under Secretary for Technology. In November 1993, former President Clinton established a National Science and Technology Council to coordinate decisionmaking in science and technology and to insure their integration at all policy levels. However, technological issues and responsibilities remain shared among many departments and agencies. This diffused focus has sometimes resulted in actions which, if not at cross purposes, may not have accounted for the impact of policies or practices in one area on other parts of the process. Technology issues involve components which operate both separately and in concert. While a diffused approach can offer varied responses to varied issues, the importance of interrelationships may be underestimated and their usefulness may suffer.

Several times, Congress has examined the idea of an industrial policy to develop a coordinated approach on issues of economic growth and industrial competitiveness. Technological advance is both one aspect of this and an altogether separate consideration. In looking at the development of an identified policy for industrial competitiveness, advocates argue that such an effort could ameliorate much of the uncertainty with which the private sector perceives future government actions. Some commentators have argued that consideration and delineation of national objectives could encourage industry to engage in more long-term planning with regard to R&D and to make decisions as to the best allocation of resources. Such a technology policy could generate greater consistency in government activities. Because technological development involves numerous risks, efforts to minimize uncertainty regarding federal programs and policies may help alleviate some of the disincentives perceived by industry.

The development of a technology policy, however, is a contentious issue. There is widespread resistance to what could be and has been called national planning, due variously to doubts as to its efficacy, to fear of adverse effects on our market system, to political beliefs about government intervention in our economic system, and to the current emphasis on short- term returns in both the political and economic arenas. Opponents of a national industrial policy may see this approach as government interference in the marketplace to "pick winners and losers." Instead, it is argued, measures that would occasion a better investment environment for industry to expand innovation-related efforts would be preferable to government decisionmaking in technological advancement.

Consideration of what constitutes government policy (both in terms of the industrial policy and technology policy) covers a broad range of ideas from laissez-faire to special government incentives to target specific high-technology, high-growth industries. Suggestions have been made for the creation of federal mechanisms to identify and support strategic industries and technologies. Various federal agencies and private sector groups have developed critical technology lists. However, others maintain that such targeting is an unwanted, and unwarranted, interference in the private sector which will cause unnecessary dislocations in the marketplace or a misallocation of resources. From their perspective, the government does not have the knowledge or expertise to make business-related decisions. Instead, they argue, the appropriate role for government is to encourage innovative activities in all industries and to keep market related decisionmaking within the business community that has ultimate responsibility for commercialization and where such decisions have traditionally been made.

The relationship between government and industry often is a major factor affecting innovation and the environment within which technological development takes place. This relationship can be adversarial, with the government acting to regulate or restrain the business community, rather than to facilitate its positive contributions to the nation. However, this may be changing as the benefits of industry/government cooperation become more apparent. There are an increasing number of areas where the traditional distinctions between public and private sector functions and responsibilities are becoming blurred. Many assumptions have been questioned, particularly in light of the increased internationalization of the U.S. economy. The business sector is no longer viewed in an exclusively domestic context; the economy of the United States is often tied to the economies of other nations. The technological superiority long held by the United States in many areas has been challenged by other industrialized countries in which economic, social, and political policies and practices foster government-industry cooperation in technological development.

A major divergence from the past was evident in the approach taken by the former Clinton Administration. Articulated in two reports issued in February 1993 (A Vision of Change for America and Technology for America's Economic Growth, A New Direction to Build Economic Strength),[2] the proposal called for a national commitment to, and a strategy for, technological advancement as part of a defined national economic policy. This detailed strategy offered a policy agenda for economic growth in the United States, of which technological development and industrial competitiveness were critical components.

In articulating a national technology policy, the approach initially recommended and subsequently followed by the Clinton Administration was multifaceted and provided a wide range of options while for the most part reflecting then current trends in congressional efforts to facilitate industrial advancement. This policy increased federal coordination and augmented direct government spending for technological development. While many past activities focused primarily on research, the new initiatives shifted the emphasis toward development of new products, processes, and services by the private sector for the commercial marketplace. In addition, a significant number of the proposals aimed to increase both government and private sector support for R&D leading to the commercialization of technology.

To facilitate technological advance, the Clinton approach focused on increasing investment; investment in research, primarily civilian research, to meet the Nation's needs in energy, environmental quality, and health; investment in the development and commercialization of new products, processes, and services for the marketplace; investment in improved manufacturing to make American goods less expensive and of better quality; investment in small, high technology businesses in light of their role in innovation and job creation; and investment in the country's infrastructure to support all these efforts. To make the most productive use of this increased investment, the Administration supported increased cooperation between all levels of government, industry, and academia to share risk, to share funding, and to utilize the strengths of each sector in reaching common goals of economic growth, productivity improvement, and maintenance of a high living standard. On November 23, 1993, President Clinton issued Executive Order 12881 establishing a National Science and Technology Council, a cabinet-level body to "coordinate science, space, and technology policies throughout the federal government."

The approach adopted by the former Administration has been questioned by recent Congresses and by the current Bush Administration. Instead, policies have appeared to support indirect strategies such as tax incentives, intellectual property protection, and antitrust laws to promote technology advancement, increased government support for basic research, and decreased direct federal funding for private sector technology activities. In the 2006 State of the Union Address, President Bush announced the "American Competitiveness Initiative" to facilitate innovation and provide "our nation's children a firm grounding in math and science." To achieve these goals, the President has called for doubling over the next 10 years the amount of federal funding for basic research, particularly in the National Science Foundation, the Office of Science in the Department of Energy, and in the core programs of the National Institute of Standards and Technology, Department of Commerce. In addition, the Initiative would increase the number of math and science teachers and make the research and experiment tax credit permanent.

Despite the continuing debate on what is the appropriate role of government and what constitutes a suitable government technology development policy, it remains an undisputed fact that what the government does or does not do affects the private sector and the marketplace. The various rules, regulations, and other activities of the government have become de facto policy as they relate to, and affect, innovation and technological advancement.

Legislative Initiatives and Current Programs

Legislative initiatives have reflected a trend toward expanding the government's role beyond traditional funding of mission-oriented R&D and basic research toward the facilitation of technological advancement to meet other critical national needs, including the economic growth that flows from new commercialization and use of technologies and techniques in the private sector. An overview of recent legislation shows federal efforts aimed at (1) encouraging industry to spend more on R&D; (2) assisting small high-technology businesses; (3) promoting joint research activities between companies; (4) fostering cooperative work between industry and universities; (5) facilitating the transfer of technology from the federal laboratories to the private sector; and (6) providing incentives for quality improvements. These efforts tend toward removing barriers to technology development in the private sector (thereby permitting market forces to operate) and providing incentives to encourage increased private sector R&D activities. While most focus primarily on research, some also involve policies and programs associated with technology development and commercialization.

Increased R&D Spending

To foster increased company spending on research, the 1981 Economic Recovery Tax Act (P.L. 97-34) mandated a temporary incremental tax credit for qualified research expenditures. The law provided a 25% tax credit for the increase in a firm's qualified research costs above the average expenditures for the previous three tax years. Qualified costs included in-house expenditures such as wages for researchers, material costs, and payments for use of equipment; 65% of corporate grants towards basic research at universities and other relevant institutions; and 65% of payments for contract research. The credit applied to research expenditures through 1985.

The Tax Reform Act of 1986 (P.L. 99-5 14) extended the research and experimentation (R&E) tax credit for another three years. However, the credit was lowered to 20% and made applicable to only 75% of a company's liability. The 1988 Tax Corrections Act (P.L. 100-647) approved a one-year extension of the research tax credit. The Omnibus Budget Reconciliation Act (P.L. 10 1-239) extended the credit through September 30, 1990 and made small start-up firms eligible for the credit. The FY1991 Budget Act (P.L. 101-508) again continued the tax credit provisions through 1992. The law expired in June 1992 when former President Bush vetoed H.R. 11 that year. However, P.L. 103-66, the Omnibus Budget Reconciliation Act of 1993, reinstated the credit through July 1995 and made it retroactive to the former expiration date. The tax credit again was allowed to expire until P.L. 104-188, the Small Business Job Protection Act, restored it from July 1, 1996 through May 31, 1997. P.L. 105-34, the Taxpayer Relief Act of 1997, extended the credit for 13 months from June 1, 1997 through June 30, 1998. Although it expired once again at the end of June, the Omnibus Consolidated Appropriations Act, P.L. 105-277, reinstated the tax credit through June 30, 1999. During the 105th Congress, various bills were introduced to make the tax credit permanent; other bills would have allowed the credit to be applied to certain collaborative research consortia. On August 5, 1999, both the House and Senate agreed to the conference report for H.R. 2488, the Financial Freedom Act, which would have extended the credit for five years through June 30, 2004. This bill also would have increased the credit rate applicable under the alternative incremental research credit by one percentage point per step. While the President vetoed the overall appropriations bill on September 23, 1999, the same provisions were included in Title V of P.L. 106-170 signed into law on December 17, 1999. P.L. 108-311 extended the research tax credit through December 31, 2005 while P.L. 109-43 2 extends the credit through the end of 2007.[3]

The Small Business Development Act (P.L. 97-2 19), as extended (P.L. 99-443), established a program to facilitate increased R&D within the small-business, high- technology community. Each federal agency with a research budget was required to set aside 1.25% of its R&D funding for grants to small firms for research in areas of interest to that agency. P.L. 102-564, which reauthorized the Small Business Innovation Research (SBIR) program, increased the set-aside over a five-year period to 2.5% by 1997. Funding is, in part, dependent on companies obtaining private sector support for the commercialization of the resulting products or processes. The authorization for the program was set to terminate October 1, 2000. However, the SBIR activity was reauthorized through September 30, 2008 by P.L. 106-554, signed into law on December 21, 2000. P.L. 102-564 also created a pilot effort, the Small Business Technology Transfer (STTR) program, to encourage firms to work with universities or federal laboratories to commercialize the results of research. This program initially was funded by a 0.15% (phased in) set-aside. Set to expire in FY1997, the STTR originally was extended for one year until P.L. 105-135 reauthorized this activity through FY2001. Subsequently, P.L. 107-50 extended the program through FY2009 and expanded the set-aside to 0.3% beginning in FY2004. Also in FY2004, the amount of individual Phase II grants increased to $750,000. (See CRS Report 96-402, Small Business Innovation Research Program, by Wendy H. Schacht.)

The Omnibus Trade and Competitiveness Act of 1988 (P.L. 100-418) created the Advanced Technology Program (ATP) at the Department of Commerce's National Institute of Standards and Technology (NIST). ATP provides seed funding, matched by private sector investment, for companies or consortia of universities, industries, and/or government laboratories to accelerate development of generic technologies with broad application across industries. The first awards were made in 1991. As of the end of 2004 (after which time no new awards have been made), 768 projects had been funded representing approximately $2.3 billion in federal dollars matched by $2.1 billion in private sector financing. About 66% of the awardees are small businesses or cooperative efforts led by such firms. A new grant competition was announced in April 2007. (For more information, see CRS Report 95-36, The Advanced Technology Program, by Wendy H. Schacht.)

Appropriations for the ATP included $35.9 million in FY1991, $47.9 million in FY1992, and $67.9 million in FY1993. FY1994 appropriations increased significantly to $199.5 million and even further in FY1995 to $431 million. However, P.L. 104-6, rescinded $90 million from this amount. The original FY1996 appropriations bill, H.R. 2076, which passed the Congress, was vetoed by President Clinton, in part, because it provided no support for ATP. The appropriations legislation finally enacted, P.L. 104-134, did fund the Advanced Technology Program at $221 million. For FY1997, the President's budget request was $345 million. However, P.L. 104-208, the Omnibus Consolidated Appropriations Act, provided $225 million for ATP, later reduced by $7 million to $218 million by P.L. 105-18. The Administration's FY1998 budget requested $276 million in funding; P.L. 105-119 appropriated $192.5 million for ATP, again at a level less than the previous year. The President's FY1 999 budget proposal included $259.9 million for this program, a 35% increase. While not providing such a large increase, P.L. 105- 277 did fund ATP at $197.5 million, 3% above the previous year. This figure reflected a $6 million rescission contained in the same law that accounted for "deobligated" funds resulting from early termination of certain projects.

In FY2000, the Clinton Administration proposed $238.7 million for ATP, an increase of 21% over the previous year. H.R. 2670, as passed by the House, provided no funding for the activity. The report to accompany the House bill stated that there was insufficient evidence "to overcome those fundamental questions about whether the program should exist in the first place." S. 1217, as passed by the Senate, would have appropriated $226.5 million for ATP. P.L. 106-113 eventually did finance the program at $142.6 million, 28% below prior year funding. The following year, the President's FY2001 budget included $175.5 million for ATP, an increase of 23% over the earlier fiscal year. Once again, the original version of the appropriations bill that passed the House did not contain any financial support for the activity. However, P.L. 106-553 provided $145.7 million in FY2001 support for ATP, 2% above the previous funding level.

For FY2002, President Bush's budget proposed suspending all funding for new ATP awards pending an evaluation of the program. In the interim, $13 million would have been provided to meet the financial commitments for on-going projects. H.R. 2500, as initially passed by the House, also did not fund new ATP grants but offered $13 million for prior commitments. The version of H.R. 2500 that originally passed the Senate provided $204.2 million for the ATP effort. P.L. 107-77 funded the program at $184.5 million, an increase of almost 27% over the previous fiscal year.

The Bush Administration's FY2003 budget request would have funded ATP at $108 million; 35% below the FY2002 appropriation level. While no relevant appropriations legislation was passed by the 107th Congress, a series of Continuing Resolutions funded the program until the 108th Congress enacted P.L. 108-7 which financed ATP at $178.8 million for FY2003 (after a mandated 0.65% across the board recision).

In its FY2004 budget, the Administration proposed to provide $17 million to cover on-going commitments to ATP; however no new projects would be funded. H.R. 2799, the FY2004 appropriations bill initially passed by the House, included no support for ATP. Subsequently incorporated into H.R. 2673, which became P.L. 108-199, the legislation funded ATP at $179.2 million (prior to a mandated 0.59% across the board rescission). As reported to the Senate from the Committee on Appropriations, S. 1585 would have financed the program at $259.6 million.

The President's FY2005 budget, as well as H.R. 4754, the Commerce, Justice, State Appropriations bill originally passed by the House, did not include any funding for ATP. As reported to the Senate from the Committee on Appropriations, S. 2809 would have provided $203 million for the program, 19% above the previous fiscal year. P.L. 108-447, the FY2005 Omnibus Appropriations Act, funded ATP at $136.5 million (after several rescissions mandated in the legislation), 20% below FY2004.

For FY2006, the Administration's budget and H.R. 2862, as originally passed by the House, again did not include funding for the Advanced Technology Program. The version of H.R. 2862 initially passed by the Senate would have provided ATP with $140 million. The final FY2006 appropriation legislation, P.L. 109-108, finances the program at $79 million (after mandated rescissions), 42% less than the last fiscal year.

The President's FY2007 budget did not include funding for ATP, nor did H.R. 5672, the FY2007 Science, State, Justice, Commerce, and Related Agencies Appropriations Act, as passed by the House on June 29, 2006 and as reported from the Senate Committee on Appropriations. While no final FY2007 appropriations legislation was enacted during the 109th Congress, ATP was funded through February 15, 2007 by a series of continuing resolutions. Passed in the 110th Congress, P.L. 110-5 provides FY2007 appropriations of $79 million for the program.

The Administration's FY2008 budget proposal does not include support for the Advanced Technology Program. H.R. 3093, as passed by the House, would fund the program at $93.1 million; S. 1745, as reported from the Senate Committee on Appropriations, would provide $100 million.

The 21st Century Competitiveness Act (H.R. 2272), as agreed to in conference, would create a new Technology Innovation Program (TIP) to replace ATP. While similar to ATP in the intent to promote high-risk R&D that would be of broad-based economic benefit to the nation, there are several differences in the operation of the new activity. Funding under TIP would be limited to small and medium-sized businesses whereas grants under ATP are available to companies regardless of size. In addition, in the Advanced Technology Program, joint ventures must include two separately owned for-profit firms and may include universities, government laboratories, and other research establishments as participants in the project but not as recipients of the grant. In the TIP initiative, a joint venture may involve two separately owned for-profit companies but may also be comprised of one small or medium-sized firm and a university (or other non-profit research organization). A single company may receive up to $2 million dollars for up to three years under ATP; under TIP, the participating company (which must be a small or medium-sized business) may receive up to $3 million for up to three years. In ATP, small and medium-sized companies are not required to cost share (large firms must provide 60% of the total cost of the project) while in TIP there is a 50% cost sharing requirement which, again, only applies to the small and medium-sized businesses that are eligible. There are no funding limits for the five-year funding available for joint ventures under ATP; the TIP limits joint venture funding to $9 million for up to five years. The Advisory Board that was created to assist in the Advanced Technology Program includes industry representatives as well as federal government personnel and representatives from other research organizations. The Advisory Board for the Technology Innovation Program would be comprised of only private sector members.

Industry-University Cooperative Efforts

The promotion of cooperative efforts among academia and industry is aimed at increasing the potential for the commercialization of technology. (For more information, see CRS Report RL3 3526, Cooperative R&D: Federal Efforts to Promote Industrial Competitiveness, by Wendy H. Schacht.) Traditionally, basic research has been performed in universities or in the federal laboratory system while the business community focuses on the manufacture or provision of products, processes, or services. Universities are especially suited to undertake basic research. Their mission is to educate and basic research is an integral part of the educational process. Universities generally are able to undertake these activities because they do not have to produce goods for the marketplace and therefore can do research not necessarily tied to the development of a commercial product or process.

Subsequent to World War II, the federal government supplanted industry as the primary source of funding for basic research in universities. It also became the principal determinant of the type and direction of the research performed in academia. This resulted in a disconnect between the university and industrial communities. The separation and isolation of the parties involved in the innovation process is thought by many observers to be a barrier to technological progress. The difficulties in moving an idea from the concept stage to a commercial product or process may be compounded when several entities are involved. Legislation to stimulate cooperative efforts among those involved in technology development has been viewed as one way to promote innovation and facilitate the international competitiveness of U.S. industry.

Several laws have attempted to encourage industry-university cooperation. Title II of the Economic Recovery Tax Act of 1981 (P.L. 97-34) provided, in part, a 25% tax credit for 65% of all company payments to universities for the performance of basic research. Firms were also permitted a larger tax deduction for charitable contributions of equipment used in scientific research at academic institutions. The Tax Reform Act of 1986 (P.L. 99-5 14) kept this latter provision, but reduced the credit for university basic research to 20% of all corporate expenditures for this over the sum of a fixed research floor plus any decrease in non-research giving.

The 1981 act also provided an increased charitable deduction for donations of new equipment by a manufacturer to an institution of higher education. This equipment must be used for research or research training for physical or biological sciences within the United States. The tax deduction is equal to the manufacturer's cost plus one-half the difference between the manufacturer's cost and the market value, as long as it does not exceed twice the cost basis. These provisions were extended through July 1995 by the Omnibus Budget Reconciliation Act of 1993, but then expired until restored by the passage of P.L. 104-188, P.L. 105-277, and P.L. 106-170 as noted above. H.R. 6111, passed by both the House and Senate during the 109th Congress and awaiting the President's signature, extends the research credit through the end of 2007.

Amendments to the patent and trademark laws contained in P.L. 96-5 17 (commonly called the "Bayh-Dole Act") also were designed to foster interaction between academia and the business community. This law provides, in part, for title to inventions made by contractors receiving federal R&D funds to be vested in the contractor if they are small businesses, universities, or not-for-profit institutions. Certain rights to the patent are reserved for the government and these organizations are required to commercialize within a predetermined and agreed upon time frame. Providing universities with patent title is expected to encourage licensing to industry where the technology can be manufactured or used thereby creating a financial return to the academic institution. University patent applications and licensing have increased significantly since this law was enacted. (See CRS Report RL32076, The Bayh-Dole Act: Selected Issues in Patent Policy and the Commercialization of Technology and CRS Report RL30320, Patent Ownership and Federal Research and Development: A Discussion on the Bayh-Dole Act and the Stevenson-Wydler Act, both by Wendy H. Schacht.)

The CREATE Act, P.L. 108-453, makes changes in the patent laws to promote cooperative research and development among universities, government, and the private sector. The bill amends section 103(c) of title 25, United States Code, such that certain actions between researchers under a joint research agreement will not preclude patentability. (For more detail see CRS Report RS2 1882, Collaborative R&D and the Cooperative Research and Technology Enhancement (CREATE) Act, by Wendy H. Schacht.)

Joint Industrial Research

Private sector investments in basic research are often costly, long term, and risky. Although not all advances in technology are the result of research, it is often the foundation of important new innovations. To encourage increased industrial involvement in research, legislation was enacted to allow for joint ventures in this arena. It is argued that cooperative research reduces risks and costs and allows for work to be performed that crosses traditional boundaries or expertise and experience. Such collaborative efforts make use of existing and support the development of new resources, facilities, knowledge, and skills.

The National Cooperative Research Act (P.L. 98-462) encourages companies to undertake joint research. The legislation clarifies the antitrust laws and requires that a "rule of reason" standard be applied in determinations of violations of these laws; cooperative research ventures are not to be judged illegal "per se." It eliminates treble damage awards for those research ventures found in violation of the antitrust laws if prior disclosure (as defined in the law) has been made. P.L. 98-462 also makes changes in the way attorney fees are awarded. Defendants can collect attorney fees in specified circumstances, including when the claim is judged frivolous, unreasonable, without foundation, or made in bad faith. However, the attorney fee award to the prevailing party may be offset if the court decides that the prevailing party conducted a portion of the litigation in a manner which was frivolous, unreasonable, without foundation, or in bad faith. These provisions were included to discourage frivolous litigation against joint research ventures without simultaneously discouraging suits of plaintiffs with valid claims. Between 1985 (when the law went into effect) and 2003, 913 joint research ventures have filed with the Department of Justice.

P.L. 103-42, the National Cooperative Production Amendments Act of 1993, amends the National Cooperative Research Act by, among other things, extending the original law's provisions to joint manufacturing ventures. These provisions are only applicable, however, to cooperative production when (1) the principal manufacturing facilities are "located in the United States or its territories, and (2) each person who controls any party to such venture ... is a United States person, or a foreign person from a country whose law accords antitrust treatment no less favorable to United States persons than to such country's domestic persons with respect to participation in joint ventures for production."

Commercialization of the Results of Federally Funded R&D

Another approach to encouraging the commercialization of technology involves the transfer of technology from federal laboratories and contractors to the private sector where commercialization can proceed. Because the federal laboratory system has extensive science and technology resources and expertise developed in pursuit of mission responsibilities, it is a potential source of new ideas and knowledge which may be used in the business community. (See CRS Report RL3 3527, Technology Transfer: Utilization of Federally Funded Research and Development, by Wendy H. Schacht for more details.)

Despite the potential offered by the resources of the federal laboratory system, however, the commercialization level of the results of federally funded R&D remained low. Studies indicated that only approximately 10% of federally owned patents were ever utilized. There are many reasons for this low level of usage, one of which is the fact that some technologies and/or patents have no market application. However, industry unfamiliarity with these technologies, the "not-invented-here" syndrome, and perhaps more significantly, the ambiguities associated with obtaining title to or exclusive license to federally owned patents also contribute to the low level of commercialization.

Over the years, several governmental efforts have been undertaken to augment industry's awareness of federal R&D resources. The Federal Laboratory Consortium for Technology Transfer was created in 1972 (from a Department of Defense program) to assist in transferring technology from the federal government to state and local governments and the private sector. To expand on the work of the Federal Laboratory Consortium, and to provide added emphasis on the commercialization of government technology, Congress passed P.L. 96-480, the Stevenson-Wydler Technology Innovation Act of 1980. Prior to this law, technology transfer was not an explicit mandate of the federal departments and agencies with the exception of the National Aeronautics and Space Administration. To provide "legitimacy" to the numerous technology activities of the government, Congress, with strong bipartisan support, enacted P.L. 96-480 which explicitly states that the federal government has the responsibility, "to ensure the full use of the results of the nation's federal investment in research and development." Section 11 of the law created a system within the federal government to identify and disseminate information and expertise on what technologies or techniques are available for transfer. Offices of Research and Technology Applications were established in each federal laboratory to distinguish technologies and ideas with potential applications in other settings.

Several amendments to the Stevenson-Wydler Technology Innovation Act have been enacted to provide additional incentives for the commercialization of technology. P.L. 99-502, the Federal Technology Transfer Act, authorizes activities designed to encourage industry, universities, and federal laboratories to work cooperatively. It also establishes incentives for federal laboratory employees to promote the commercialization of the results of federally funded research and development. The law amends P.L. 96-480 to allow government-owned, government-operated laboratories to enter into cooperative R&D agreements (CRADAs) with universities and the private sector. This authority is extended to government-owned, contractor-operated laboratories by the Department of Defense FY1990 Authorization Act, P.L. 101-189. (See CRS Report 95-150, Cooperative Research and Development Agreements (CRADAs), by Wendy Schacht.) Companies, regardless of size, are allowed to retain title to inventions resulting from research performed under cooperative agreements. The federal government retains a royalty-free license to use these patents. The Technology Transfer Improvements and Advancement Act (P.L. 104-113), clarifies the dispensation of intellectual property rights under CRADAs to facilitate the implementation of these cooperative efforts. The Federal Laboratory Consortium is given a legislative mandate to assist in the coordination of technology transfer. To further promote the use of the results of federal R&D, certain agencies are mandated to create a cash awards program and a royalty sharing activity for federal scientists, engineers, and technicians in recognition of efforts toward commercialization of this federally developed technology. These efforts are facilitated by a provision of the National Defense Authorization Act for FY1 991 (P.L. 101-510), which amends the Stevenson-Wydler Technology Innovation Act to allow government agencies and laboratories to develop partnership intermediary programs to augment the transfer of laboratory technology to the small business sector.

Amendments to the Patent and Trademark law contained in Title V of P.L. 98- 620 made changes which are designed to improve the transfer of technology from the federal laboratories — especially those operated by contractors — to the private sector and increase the chances of successful commercialization of these technologies. This law permits the contractor at government-owned, contractor-operated laboratories (GOCOs) to make decisions at the laboratory level as to the granting of licenses for subject inventions. This has the potential of effecting greater interaction between laboratories and industry in the transfer of technology. Royalties on these inventions are also permitted to go back to the laboratory contractor to be used for additional R&D, awards to individual laboratory inventors, or education. While there is a cap on the amount of the royalty returning directly to the lab in order not to disrupt the agency's mission requirements and congressionally mandated R&D agenda, the establishment of discretionary funds gives contractor-operated laboratories added incentive to encourage technology transfer.

Under P.L. 98-620, private companies, regardless of size, are allowed to obtain exclusive licenses for the life of the patent. Prior restrictions allowed large firms use of exclusive license for only 5 of the 17 years (now 20 years) of the life of the patent. This was expected to encourage improved technology transfer from the federal laboratories or the universities (in the case of university operated GOCOs) to large corporations which often have the resources necessary for development and commercialization activities. In addition, the law permits GOCOs (those operated by universities or nonprofit institutions) to retain title to inventions made in the laboratory within certain defined limitations. Those laboratories operated by large companies are not included in this provision.

P.L. 106-404, the Technology Transfer Commercialization Act, altered practices concerning patents held by the government to make it easier for federal agencies to license such inventions. The law amends the Stevenson-Wydler

Technology Innovation Act and the Bayh-Dole Act to decrease the time delays associated with obtaining an exclusive or partially exclusive license. Previously, agencies were required to publicize the availability of technologies for three months using the Federal Register and then provide an additional 60 day notice of intent to license by an interested company. Under this legislation, the time period was shortened to 15 days in recognition of the ability of the internet to offer widespread notification and the necessity of time constraints faced by industry in commercialization activities. Certain rights are retained by the government. The bill also allows licenses for existing government-owned inventions to be included in CRADAs.

The Omnibus Trade and Competitiveness Act (P.L. 100-4 18) mandated the creation of a program of regional centers to assist small manufacturing companies to use knowledge and technology developed under the auspices of the National Institute of Standards and Technology and other federal agencies. Federal funding for the centers is matched by non-federal sources including state and local governments and industry. Originally, seven Regional Centers for the Transfer of Manufacturing Technology were selected. The initial program was expanded in 1994 to create the Manufacturing Extension Partnership (MEP) to meet new and growing needs of the community. In a more varied approach, the Partnership involves both large centers and smaller, more dispersed organizations sometimes affiliated with larger centers as well as the NIST State Technology Extension Program which provides states with grants to develop the infrastructure necessary to transfer technology from the federal government to the private sector (an effort which was also mandated by P.L. 100-418) and a program which electronically ties the disparate parties together along with other federal, state, local, and academic technology transfer organizations. There are now centers in all 50 states and Puerto Rico. Since the manufacturing extension activity was created in 1989, awards made by NIST have resulted in the creation of approximately 350 regional offices. [It should be noted that the Department of Defense also funded 36 centers through its Technology Reinvestment Project (TRP) in FY1994 and FY1995. When the TRP was terminated, NIST took over support for 20 of these programs in FY1996 and funded the remaining efforts during FY1997.]

Funding for this program was $11.9 million in FY1991, $15.1 million in FY1992, and $16.9 million in FY1993. In FY1994 support for the expanded Manufacturing Technology Partnerships was $30.3 million. The following fiscal year, P.L. 103-3 17 appropriated $90.6 million for this effort, although P.L. 104-19 rescinded $16.3 million from this amount. While the original FY1 996 appropriations bill, H.R. 2076, was vetoed by the President, the $80 million funding for MEP was retained in the final legislation, P.L. 104-134. The

President's FY1997 budget request was $105 million; P.L. 104-208 appropriated $95 million for manufacturing extension while temporarily lifting the six-year limit on federal support for individual centers. For FY1998, the Administration requested funding of $123 million. The FY1998 appropriations bill, P.L. 105-119, financed the MEP program at $113.5 million. This law also permitted government funding, at one-third the centers total annual cost, to continue for additional periods of one year over the original six-year limit, if a positive evaluation is received. The President's FY1999 budget included $106.8 million for the MEP, a 6% decrease from current funding. The Omnibus Consolidated Appropriations Act, P.L. 105-277, appropriated the $106.8 million. The decrease in funding reflected a reduced federal financial commitment as the centers mature, not a decrease in program support. In addition, the Technology Administration Act of 1998, P.L. 105-309, permits the federal government to fund centers at one-third the cost after the six years if a positive, independent evaluation is made every two years.

For FY2000, the Clinton Administration requested $99.8 million in support for MEP. Again, the lower federal share indicated a smaller statutory portion required of the government. S. 1217, as passed by the Senate, would have appropriated $109.8 million for the Manufacturing Extension Partnership, an increase of 3% over FY1 999. H.R. 2670, as passed initially by the House, would have appropriated $99.8 million for this activity. The version of the H.R. 2670 passed by both House and Senate provided FY2000 appropriations of $104.8 million. While the President vetoed that bill, the legislation that was ultimately enacted, P.L. 106-113, appropriated $104.2 million after a mandated rescission. The President's FY2001 budget requested $114.1 million for the Partnership, an increase of almost 9% over the earlier fiscal year. P.L. 106-553 appropriated $105.1 million.

The FY2002 Bush Administration budget proposed providing $106.3 million for MEP. H.R. 2500, as originally passed by the House, would have funded MEP at $106.5 million. The initial version of H.R. 2500 passed by the Senate would have provided $105.1 million for the program. The final legislation, P.L. 107-77 funded the Partnership at $106.5 million.

For FY2003, the Administration's budget included an 89% decrease in support for MEP. According to the budget document, "consistent with the program's original design, the President's budget recommends that all centers with more than six years experience operate without federal contribution." A number of Continuing Resolutions supported the Partnership at FY2002 levels until the 108th Congress enacted P.L. 108-7 which appropriated $105.9 million for MEP in FY2003 (after a mandated recision).

The President's FY2004 budget requested $12.6 million for MEP to finance only those centers that have not reached six years of federal support. H.R. 2799, as initially passed by the House, would have appropriated $39.6 million for the Partnership. This bill was subsequently incorporated into H.R. 2673, which became P.L. 108-199, the FY2004 Consolidated Appropriations Act. This legislation financed MEP at $38.7 million after a mandated rescission. S. 1585, reported to the Senate by the Committee on Appropriations, would have funded the program at $106.6 million.

The Administration proposed funding MEP at $39.2 million in FY2005. H.R. 4754, as originally passed by the House, would have appropriated $106 million for this program. As reported by the Senate Committee on Appropriations, S. 2809 would have provided $112 million for MEP to "fully fund" existing centers and provide assistance to small and rural states. P.L. 108-447, supported manufacturing extension at $107.5 million (after several mandated rescissions included in the legislation).

For FY2006, the President's budget requested $46.8 million for the Manufacturing Extension Partnership, 56% below funding for the current fiscal year. H.R. 2862, as originally passed by both the House and the Senate, would have provided $106 million for the program. The final appropriation included in P.L. 109-108 was $104.6 million (after mandated rescissions, but not including a rescission from unobligated balances).

The Administration's FY2007 budget included $46.3 million for MEP. The FY2007 appropriations bill passed by the House, H.R. 5672, funded the program at $92 million. The version of H.R. 5672 reported from the Senate Committee on Appropriations provided $106 million for MEP. No final FY2007 appropriations legislation was enacted during the 109th Congress; however, the Partnership program was funded through February 15, 2007 by a series of continuing resolutions. Passed by the current Congress, P.L. 110-5 finances MEP at $104.6 million in FY2007.

The President's FY2008 budget proposal includes $46.3 million for manufacturing extension, a significant decrease from the current fiscal year. H.R. 3093, as passed by the House, would fund MEP at $108.8 million, while S. 1745, as reported would provide $110 million in FY2008.

In the current Congress, two bills passed by the House (H.R. 1868 and H.R. 2272) authorize a new program of partnerships between industry and other educational or research institutions to develop new manufacturing processes, techniques, or materials. In addition, a manufacturing fellowship program would be created with stipends available for post-doctoral work at NIST. These activities differ from the established MEP effort where no new manufacturing research is conducted as existing manufacturing technology is applied to the needs of small and medium-sized firms. (For additional information see CRS Report 97-104, Manufacturing Extension Partnership Program: An Overview, by Wendy Schacht.)

Different Approach?

As indicated above, the laws affecting the R&D environment have included both direct and indirect measures to facilitate technological innovation. In general, direct measures are those which involve budget outlays and the provision of services by government agencies. Indirect measures include financial incentives and legal changes (e.g., liability or regulatory reform; new antitrust arrangements). Supporters of indirect approaches argue that the market is superior to government in deciding which technologies are worthy of investment. Mechanisms that enhance the market's opportunities and abilities to make such choices are preferred. Advocates further state that dependency on agency discretion to assist one technology in preference to another will inevitably be subjected to political pressures from entrenched interests. Proponents of direct government assistance maintain, conversely, that indirect methods can be wasteful and ineffective and that they can compromise other goals of public policy in the hope of stimulating innovative performance. Advocates of direct approaches argue that it is important to put the country's scarce resources to work on those technologies that have the greatest promise as determined by industry and supported by its willingness to match federal funding.

In the past, while Republicans tended to prefer reliance on free market investment, competition, and indirect support by government, participants in the debates generally did not make definite (or exclusionary) choices between the two approaches, nor consistently favor one over the other. For example, some proponents of a stronger direct role for the government in innovation are also supporters of enhanced tax preferences for R&D spending, an indirect mechanism. Opponents of direct federal support for specific projects (e.g., SEMATECH, flat panel displays) may nevertheless back similar activities focused on more general areas such as manufacturing or information technology. However, beginning with the 104th Congress, legislators directed many of their efforts toward eliminating or curtailing some of the programs that previously had enjoyed bipartisan support. Initiatives to terminate the Advanced Technology Program, funding for flat panel displays, and agricultural extension reflected concern about the role of government in developing commercial technologies.

database. Introduced April 17, 2007; referred to the House Committee on Science and Technology. Reported, amended, to the House on April 30, 2007. Passed by the House on May 3, 2007 and referred to the Senate Committee on Commerce, Science, and Transportation on May 7, 2007.

H.R. 2272 (Gordon)

21st Century Competitiveness Act of 2007. Title IV authorizes funding for the National Institute of Standards and Technology (NIST) through 2010 and creates several new manufacturing R&D programs in that organization. Funding for the Scientific and Technical Research and Services account within NIST is authorized at $471 million for FY2008, $498 million for FY2009, and $538 for FY2010. Authorizations for the Malcolm Baldrige National Quality Award Program would include $7.9 million in FY2008, $8.1 million in FY2009, and $8.3 million in FY20 10. Support for construction and maintenance would be authorized at $94 million for FY2008, $86 million for FY2009, and $50 million for FY20 10. Authorization of appropriations for Industrial Technology Services programs within NIST would include $223 million ($110 million for the Technology Innovation Program (TIP) and $113 million for MEP) for FY2008, $264 million ($142 million for TIP and $122 million for MEP) for FY2009, and $282 million ($150 million for TIP and $132 million for MEP) for FY20 10. Among the new programs established within NIST would be a MEP Advisory Board, a Technology Innovation Program (to replace the Advanced Technology Program), collaborative manufacturing research pilot grants, a manufacturing fellowship program, and a manufacturing research database. Introduced on May 10, 2007; referred to the House Committee on Science and Technology. Passed House on May 21, 2007 and received in the Senate on May 22, 2007. Placed on Senate Legislative Calendar under General Orders. Senate struck out all after the Enacting Clause and substituted the language of S. 761. Passed Senate, with the amendment, on July 19, 2007. Conference held on July 31, 2007. Conference report filed on August 1, 2007.

H.R. 3093 (Mollohan)

Makes appropriations for the Departments of Commerce and Justice, and science and related agencies for FY2008. Provides $108.8 million for the Manufacturing Extension Program and $93.1 million for the Advanced

Technology Program, among other things. Introduced July 19, 2007; reported from the House Committee on Appropriations as an original measure. Passed House, amended, July 26, 2007.

S. 41 (Baucus)

Research Competitiveness Act of 2007. Amends the Internal Revenue Code to make the research and experimentation tax credit permanent. Among other things, this bill would allow the issuance of tax exempt facility bonds for research park facilities used for research and experimentation. Introduced January 4, 2007; referred to the Senate Committee on Finance.

S. 69 (Kohl)

Authorizes appropriations for the Manufacturing Extension Partnership through 2012, among other things. Introduced January 4, 2007; referred to the Senate Committee on the Judiciary. Discharged from the Senate Committee on the Judiciary by unanimous consent on January 22, 2007 and referred to the Senate Committee on Commerce, Science, and Transportation the same day.

S. 592 (Collins)

GoMe Act. Extends the research tax credit through 2012, among other things. Introduced February 14, 2007; referred to the Senate Committee on Finance.

S. 761 (Reid)

America Creating Opportunities to Meaningfully Promote Excellence in Technology, Education, and Science Act. Mandates a National Science and Technology Summit to access the state of U.S. science and technology. Requires a study on barriers to innovation and creates a National Innovation Medal and a President's Council on Innovation and Competitiveness. Authorizes appropriations for NIST through 2011 including $704 million for FY2008, $774 million for FY2009, $851 million for FY2010, and $937 million for FY201 1. Requires that NIST establish an Innovation Acceleration Research Program to facilitate manufacturing innovation, among other things. Introduced March 5, 2007; placed on Senate Legislative Calendar under General Orders March 6, 2007. Passed Senate, amended, on April 25, 2007. Received in the House on April 30, 2007. Senate incorporated this measure in H.R. 2272 as an amendment on July 19, 2007.

S. 833 (Coleman)

COMPETE Act of 2007. Makes the research and experimentation tax credit permanent, among other things. Introduced March 9, 2007; referred to the Senate Committee on Finance.

S. 1745 (Mikulski)

Departments of Commerce and Justice, Science, and Related Agencies Appropriations Act, 2008. Funds MEP at $110 million and ATP at $100 million (with a $10 million rescission from these two programs), among other things. Introduced June 29, 2007; reported from the Senate Committee on Appropriations as an original measure.

REFERENCES

[1] Edwin Mansfield, "Social Returns From R&D: Findings, Methods, and Limitations," *Research/Technology Management*, November-December 1991, 24. See also Charles I. Jones and John C. Williams, "Measuring the Social Return to R&D," *Quarterly Journal of Economics*, November 1998, 1119 and Richard R. Nelson and Paul M. Romer, "Science, Economic Growth, and Public Policy," in Bruce R. Smith and Claude E. Barfield, eds. *Technology, R&D, and the Economy*, (Washington, The Brookings Institution and the American Enterprise Institute, Washington, 1996), 57.
[2] Available from author.
[3] For additional information see CRS Report RL3 1181, *Research Tax Credit: Current Status and Selected Issues for Congress*, by Gary Guenther.

Chapter 2

COOPERATIVE R&D: FEDERAL EFFORTS TO PROMOTE INDUSTRIAL COMPETITIVENESS*

Wendy H. Schacht

ABSTRACT

In response to the foreign challenge in the global marketplace, the United States Congress has explored ways to stimulate technological advancement in the private sector. The government has supported various efforts to promote cooperative research and development activities among industry, universities, and the federal R&D establishment designed to increase the competitiveness of American industry and to encourage the generation of new products, processes, and services.

Collaborative ventures are intended to accommodate the strengths and responsibilities of all sectors involved in innovation and technology development. Academia, industry, and government often have complementary functions. Joint projects allow for the sharing of costs, risks, facilities, and expertise.

Cooperative activity covers various institutional and legal arrangements including industry-industry, industry-university, and industry-government efforts. Proponents of joint ventures argue that they permit work to be done that is too expensive for one company to support and allow for R&D that crosses traditional boundaries of expertise and experience. Such

---

* Excerpted from CRS Report RL33526, dated April 27, 2007.

arrangements make use of existing, and support the development of new, resources, facilities, knowledge, and skills. Opponents argue that these endeavors dampen competition necessary for innovation.

Federal efforts to encourage cooperative activities include the National Cooperative Research Act; the National Cooperative Production Act; tax changes permitting credits for industry payments to universities for R&D and deductions for contributions of equipment used in academic research; and amendments to the patent laws vesting title to inventions made under federal funding in universities. Technology transfer from the government to the private sector is facilitated by several laws. In addition, there are various ongoing cooperative programs supported by various federal departments and agencies.

Given the increased popularity of cooperative programs, questions might be raised as to whether they are meeting expectations. Among the issues before Congress are whether joint ventures contribute to industrial competitiveness and what role, if any, the government has in facilitating such arrangements.

MOST RECENT DEVELOPMENTS

Congressional initiatives over the past 25 years have promoted cooperative research and development among industry, universities, and the federal R&D establishment. This is evident in legislation creating technology transfer mechanisms as well as in support for two extramural programs of the National Institute of Standards and Technology (NIST): the Advanced Technology Program (ATP) which provides seed funding, matched by private sector investment, to companies or consortia for the development of generic technologies that have broad application across industrial sectors, and the Manufacturing Extension Partnership (MEP) which offers technical assistance to small and medium-sized firms through regional centers in conjunction with state or local government, universities, or the private sector.

P.L. 110-5, enacted during the 110th Congress, provides FY2007 appropriations of $79 million for ATP and $104.6 million for MEP. The President's FY2008 budget requests a significant decrease in support for manufacturing extension to $46.3 million and includes no funding for ATP. Also introduced in the 110th Congress, H.R. 255, the Manufacturing Technology Competitiveness Act of 2007, would establish several new manufacturing technology programs for small and medium-sized firms. Appropriations for MEP through 2012 would be authorized by S. 69. The Technology Innovation and Manufacturing Stimulation Act of 2007, H.R. 1868, authorizes funding for NIST through 2010 and creates several new manufacturing R&D programs in that organization. Several bills introduced in the current Congress are consistent with parts of the "American Competitiveness

Initiative" announced by the President in the 2006 State of the Union Address that proposed several innovation-related activities including increased basic research funding, making permanent the research and experimentation tax credit (which was extended through the end of 2007 by P.L. 109-432), and improved math and science education. S. 833, the Competitiveness Through Education, Technology, and Enterprise Act of 2007, would make the research tax credit permanent as does S. 41, the Research Competitiveness Act of 2007, and H.R. 1712, the Research and Development Tax Credit Act of 2007, which also create tax exempt facility bonds for the development of research park facilities, among other things. S. 592 extends the research credit through 2012. H.R. 85, the Energy Technology Transfer Act, establishes a program of grants to non-profit institutions, state and local governments, cooperative extension services, or universities to transfer energy efficient methods and technologies. H.R. 363, the Sowing the Seeds Through Science and Engineering Research Act, as passed by the House would authorize a Presidential Innovation Award, among other things. S. 761, the America Creating Opportunities to Meaningfully Promote Excellence in Technology, Education, and Science Act, passed by the Senate, provides for the creation of several programs, studies, and initiatives designed to improve U.S. innovation and competitiveness, among other things.

RATIONALE

In response to concerns over competition from foreign firms, the U.S. Congress has increasingly looked for ways the federal government can stimulate technological innovation in the private sector. This technological advancement is critical in that it contributes to economic growth and long term increases in our standard of living. New technologies can create new industries and new jobs; expand the types and geographic distribution of services; and reduce production costs by making more efficient use of resources. The development and application of technology also plays a major role in determining patterns of international trade by affecting the comparative advantages of industrial sectors. Since technological progress is not necessarily determined by economic conditions, it can have effects on trade independent of shifts in macroeconomic factors that may affect the marketplace.

Joint ventures are an attempt to facilitate technological advancement within the industrial community. Academia, industry, and government can play complementary roles in technology development. While opponents argue that cooperative ventures stifle competition, proponents assert that they are designed to accommodate the strengths and responsibilities of these sectors. Collaborative projects attempt to utilize and integrate what the participants do best and to direct these efforts toward the goal of generating new goods, processes, and services for the marketplace. They allow for shared costs, shared risks, shared facilities, and shared expertise.

The lexicon of current cooperative activity covers various different institutional and legal arrangements. These ventures might include industry-industry joint projects involving the creation of a new entity to undertake research, the reassignment of researchers to a new effort, and/or hiring new personnel. Collaborative industry-university efforts may revolve around activities in which industry supports centers (sometimes cross-disciplinary) for research at universities, funds individual research projects, and/or exchanges personnel. Cooperative activities with the federal government might include projects that use federal facilities and researchers, federal funding for industry-industry or industry-university efforts, or financial support for centers of excellence at universities to which the private sector has access.

There are many different types of cooperative arrangements. The flexibility associated with this concept can allow for the development of institutional and organizational plans tailored to the specific needs of the particular project. Issues of patent ownership, disclosure of information, licensing, and antitrust are to be resolved on an individual basis within the general guidelines established by law governing joint ventures.

Collaborative ventures can be structured either "horizontally" or "vertically." The former involves efforts in which companies work together to perform research and then use the results of this research within their individual organizations. The latter involves activities where researchers, producers, and users work together. Both approaches are seen as ways to address some of the perceived obstacles to the competitiveness of American firms in the marketplace.

Joint Industrial Research

Traditionally, the federal government has funded research and development to meet mission requirements; in areas where the government is the primary user of the results; and/or where there is an identified need for R&D not being performed in the private sector. Most government support is for basic research which is often long-term and highly risky for individual companies; yet research can be the foundation for breakthrough achievements which can revolutionize the marketplace. Studies have shown that inventions based on R&D are the more important ones. However, the societal benefits of research tend to be greater than those that can be captured by the firm performing the work. Thus the rationale for federal funding of research in industry.

The major emphasis of legislative activity has been on augmenting research in the industrial community. This focus is reflected in efforts to encourage companies to undertake cooperative research arrangements and expand the opportunities available for increases in research activities. Collaboration permits work to be done which is too expensive for one company to fund and also allows for R&D that crosses traditional boundaries of expertise and experience. A joint venture makes use of existing, and supports development of new resources, facilities, knowledge, and skills.

The concentration on increased research as a prelude to increased technological advancement was based upon the "pipeline model" of innovation. This process was understood to be a series of distinct steps from an idea through product development, engineering, testing, and commercialization to a marketable product, process, or service. Thus increases at the beginning of the pipeline — in research — were expected to result in analogous increases in innovation at the end. However, this model is no longer considered valid. Innovation is rarely a linear process and new technologies and techniques often occur that do not require basic or applied research or development. Most innovations are actually incremental improvements to existing products and processes. In some areas, particularly biotechnology, research is closer to a commercial product than this conception would indicate. In others, the differentiation between basic and applied research is artificial. The critical factor is the commercialization of the technology. Economic benefits accrue only when a technology or technique is brought to the marketplace where it can be sold to generate income and/or applied to increase productivity.

In the recent past, it was increasingly common to find that foreign companies were commercializing the results of U.S. funded research at a faster pace than American firms. In the rapidly changing technological environment, the speed at which a product, process, or service is brought to the marketplace is often a crucial factor in its competitiveness. The recognition that more than research needs to be done has lead to other approaches at cooperative efforts aimed at expediting the commercialization of the results of the American R&D endeavor. These include industry-university joint activities, use of the federal laboratory system by industry, and industry-industry development efforts where manufacturers, suppliers, and users work together.

Industry-University Cooperative Efforts

Industry-university cooperation in R&D is one important mechanism intended to facilitate technological innovation. Traditionally, universities perform much of the basic research integral to certain technological advancements. They are generally able to undertake fundamental research because it is part of the educational process and because they do not have to produce for the marketplace. The risks attached to work in this setting are fewer than those in industry where companies must earn profits. Universities also educate and train the scientists, engineers, and managers employed by companies.

Academic institutions do not have the commercialization capacity available in industry and necessary to translate the results of research into products and processes that can be sold in the marketplace. Thus, if the work performed in the academic environment is to be integrated into goods and services, a mechanism to link the two sectors must be available. Prior to World War II, industry was the primary source of funding for basic research in universities. This financial support helped shape priorities and build relationships. However, after the war the federal government supplanted industry as the major financial contributor and became the principal determinant of the type and direction of the research performed in academic institutions. This situation resulted in a disconnect between the university and industrial communities. Because industry and not the government is responsible for commercialization, the difficulties in moving an idea from the research stage to a marketable product or process appear to have been compounded.

Efforts to encourage increased collaboration between the academic and industrial sectors might be expected to augment the contribution of both parties to technological advancement. Company support for research within the university provides additional funds and information on the concerns and direction of industry. For many companies, access to expertise and facilities outside of the firm expands or complements available internal resources. Yet, such cooperation should not necessarily be seen as a panacea. Oftentimes, collaborative ventures fail because of various factors including conflicting goals, differing research cultures, and financial disagreements.

Federal Laboratory-Industry Interaction

The federal government can share its extensive facilities, expertise, knowledge, and new technologies with partners in a cooperative venture. In certain cases, the government laboratories have scientists and engineers with experience and skills, as well as equipment, not available elsewhere. The government also has a vested interest in technology development. It does not have the mandate or resources to manufacture goods but has a stake in the availability of products and processes to meet mission requirements. In addition, technological advancement contributes to the economic growth vital to the health and security of the nation.

Collaboration between government laboratories and industry is not, however, just a one way street. In several technological areas, particularly electronics and computer software, the private sector is more advanced in technologies important to the national defense and welfare of this country. Interaction with industry offers federal scientists and engineers valuable information to be used within the government R&D enterprise.

FEDERAL INITIATIVES IN COOPERATIVE R&D

The cooperative venture concept is not new. In the early 1970s, the National Science Foundation established its Industry-University Cooperative Research Centers program. The Electric Power Research Institute, a research organization supported by the electric power utilities, has been in operation since 1973. In the private sector, the Microelectronics and Computer Technology Corporation (MCC), which performs research for its member firms, and the Semiconductor Research Corporation (SRC), which funds research in universities, were created in the early 1980s. The difference today is the number of projects and the scope of legislative activity designed to promote cooperative ventures.

Faced with pressures from foreign competition, the government's interest appears to be expanding beyond that of funding R&D, to meeting other critical national needs including the economic growth that flows from new commercialization in the private sector. While acknowledging that the commercialization of technology is the responsibility of the business community, in the past several years the government has attempted to stimulate innovation and technological advancement in industry. These activities often involve the removal of barriers to technology development in the private sector, thereby permitting market forces to operate and the provision of incentives to encourage increased innovation related efforts in industry. Cooperative R&D efforts are a part of both these trends.

The National Cooperative Research Act (P.L. 98-462) is designed to encourage companies to undertake joint research which is typically long-term, risky, and often too expensive for one company to finance. This legislation clarifies the antitrust laws and requires that the "rule of reason" standard be applied in determinations of violations of these laws; that cooperative research ventures are not to be judged illegal "per se". It also eliminates treble damage awards for those research ventures found in violation of the antitrust laws if prior disclosure (as defined in the law) has been made. In addition, P.L. 98-462 makes some changes in the way attorney fee awards are made to discourage frivolous litigation against joint research ventures without simultaneously discouraging suits of plaintiffs with valid claims. Between 1985 (when the law went into effect) and 2003, over 900 joint ventures have filed with the Justice Department.[1]

P.L. 103-42, the National Cooperative Production Amendments Act of 1993, amends the National Cooperative Research Act by, among other things, extending the original law's provisions to joint manufacturing ventures. These provisions are only applicable, however, to cooperative production when the principal manufacturing facilities are "located in the United States or its territories, and each person who controls any party to such venture ... is a United States person, or a foreign person from a country whose law accords antitrust treatment no less favorable to United States persons than to such country's domestic persons with respect to participation in joint ventures for production."

The Omnibus Trade and Competitiveness Act of 1988 (P.L. 100-418) created the Advanced Technology Program (ATP) at the Department of Commerce's National Institute of Standards and Technology. This program provides seed funding, matched by private sector investment, to companies or consortia comprised of universities, companies, and/or government laboratories for the development of generic technologies that have broad application across industrial sectors. As of the end of 2004 (when the last new grant was issued), 768 projects have been funded representing approximately $2.3 billion in federal financing matched by $2.1 billion in financing from the private sector. Of these projects, approximately 30% were or are joint ventures. Eleven initial R&D programs were selected for funding, almost half of which involved consortia. Twenty-seven awards were made to programs in the second year; approximately one-third were *consortia*. In December 1992, 21 new ATP awards were made, including three joint ventures. Thirty additional projects were funded in 1993, and, in October 1994, 41 awards were made in four key technology areas: information infrastructure for healthcare; tools for DNA diagnostics; component-based software; and computer-integrated manufacturing for electronics. Fourteen are cooperative efforts. In November 1994, 47 additional awards were made in the general competition and in the area of manufacturing composite structures. Twenty-four involve collaborative R&D. Of the 24 awards announced on July 13, 1995, 35% of the projects in the general competition were joint ventures and 29% in the focused competition. The following month 21 additional awards were made of which 9 were cooperative efforts. In early September, another 44 grants were awarded including 19 joint ventures. Later in that month, 10 more awards were made of which three were to cooperative efforts. On January 25, 1996, an additional four projects received awards; three involved multiple firms. In March 1997, NIST announced that it would fund 8 new proposals from the FY1 996 general competition of which 2 were collaborative projects. Sixty-four awards were made in October 1997; 15 involving multiple companies. In October 1998, NIST awarded funding for 79 new projects involving more than 150 companies, 11 universities, and several federal laboratories. This reflects changes in the ATP selection criteria designed to encourage large companies to participate in joint ventures with small firms and academic institutions. Thirty-seven awards for FY1999 were made on October 7, 1999. Of these, 27 are either joint ventures or involve additional organizations working as subcontractors. In FY2001, 13 of the 59 grants involved collaborative projects while in FY2002, 10 of the 61 awards went to joint ventures. Of the 16 awards made in July 2003, 3 were for collaborative projects. In September of 2003, 44 awards were made of which 9 were joint ventures. An additional 32 awards were made in 2004, seven involving cooperative activities. (For more information, see CRS Report 95-36, *The Advanced Technology Program*, by Wendy H. Schacht.)

Appropriations for the Advanced Technology Program were $35.9 million in FY1991, $47.9 million in FY1992, and $67.9 million in FY1993. FY1994 appropriations expanded significantly to $199.5 million and even further to $431 million in FY1995. However, P.L. 104-6, the DOD Emergency Supplemental Appropriations and Rescissions Act, rescinded $90 million of this amount. The Clinton Administration's FY1996 budget request for ATP was $490.9 million. The original appropriations bill, H.R. 2076, which passed the Congress but was vetoed by the President, provided no financing for ATP. The final appropriations legislation, P.L. 104-134, funded the Advanced Technology Program at $221 million for FY1 996. The following year, FY1 997, the Omnibus Consolidated Appropriations Act (P.L. 104- 208) provided support levels of $225 million, but $7 million was rescinded by P.L. 105-18. P.L. 105-119 funded ATP at $192.5 million in FY1998. The President's FY1 999 budget included $259.9 million for this program, an increase of 35%. However, P.L. 105-277, the Omnibus Consolidated Appropriations Act, funded ATP at $197.5 million, 3% above the previous year. This figure reflected a $6 million rescission to account for "deobligated" funds resulting from prior projects that had been terminated early.

In the FY2000 budget, the Clinton Administration requested $238.7 million for ATP, an increase of 21% over FY1999. Yet H.R. 2670, as originally passed by the House, contained no appropriated funding for ATP. The report accompanying the

House bill stated that "... the program has not produced a body of evidence to overcome those fundamental questions about whether the program should exist in the first place." S. 1217, as initially passed by the Senate, would have appropriated $226.5 million, 15% more than the previous year. P.L. 106-113, the final FY2000 appropriations legislation, provided the Advanced Technology Program with $142.6 million, financing that was 28% below the level of the previous year. For FY2001, the President requested ATP funding of $175.5 million, an increase of 23% over prior year funding. The original appropriations bill, as passed by the House, again provided no support for the program. However, P.L. 106-553 did fund ATP at $145.7 million for FY2001, 2% above the previous fiscal year.

The Bush Administration's FY2002 budget proposed suspending all funding for new ATP awards pending an evaluation of the program. However, $13 million would have been provided to meet financial commitments for on-going projects. H.R. 2500, as first passed by the House, provided no support for new ATP projects but did include $13 million to fund prior year commitments. The original Senate-passed version of H.R. 2500 would have funded the program at $204.2 million. The final legislation, P.L. 107-77, financed ATP at $184.5 million, a 27% increase over FY2001.

In the FY2003 budget, the President requested $108 million for the Advanced Technology Program. This figure was 35% below the FY2002 appropriation. A number of Continuing Resolutions supported the program at FY2002 levels until the 108th Congress passed P.L. 108-7 which appropriated $178.8 million in FY2003 (after a 0.65% across the board mandated by the legislation).

The Administration's FY2004 budget included $27 million for ATP to cover on-going commitments; no new projects would be funded. H.R. 2799, the appropriations bill initially passed by the House, contained no funding for ATP. As reported to the Senate from the Committee on Appropriations, S. 1585 would have provided $259.6 million for ATP. P.L. 108-199, the FY2004 Consolidated Appropriations Act, financed the program at $170.5 million (after a mandated rescission).

For FY2005, the President's budget proposal, as well as H.R. 4754, the FY2005 appropriations bill originally passed by the House, did not include funding for ATP. As reported to the Senate by the Committee on Appropriations, S. 2809 would have financed the program at $203 million, an increase of 19% over the previous fiscal year. The FY2005 Omnibus Appropriations Act, P.L. 108-447, provided ATP with $136.5 million (after several rescissions mandated in the legislation), 20% less than FY2004.

President Bush's FY2006 budget request, as well as the version of H.R. 2862 initially passed by the House, did not include support for ATP. H.R. 2862, as originally passed by the Senate, would have funded the program at $140 million. The final FY2006 appropriations legislation, P.L. 109-108, provides $79 million for the program (after mandated rescissions), 42% below the previous fiscal year.

The Administration's FY2007 budget did not include funding for ATP, nor did H.R. 5672, the FY2007 Science, State, Justice, Commerce, and Related Agencies Appropriations Act, as passed by the House on June 29, 2006 and as reported from the Senate Committee on Appropriations. While no final FY2007 appropriations legislation was enacted during the 109th Congress, a series of continuing resolutions finances ATP at FY2006 levels through February 15, 2007 when the 1 10th Congress passed P.L. 110-5 which appropriated $79 million for the program. The President's FY2008 budget request again does not include support for this program.

Several laws have attempted to facilitate industry-university cooperation. Title II of the Economic Recovery Tax Act of 1981 (P.L. 97-34) provided, in part, a temporary 25% tax credit for 65% of all company payments to universities for the performance of basic research. Firms were also permitted a larger tax deduction for charitable contributions of equipment used in scientific research at academic institutions. The Tax Reform Act of 1986 (P.L. 99-5 14) kept this latter provision, but reduced the credit for university basic research to 20% of all corporate expenditures for this work over the sum of a fixed research floor plus any decrease in non-research giving.

The 1981 Act also provided an increased charitable deduction for donations of new equipment by a manufacturer to an institution of higher education. This equipment must be used for research or training for physical or biological sciences within the United States. The tax deduction was equal to the manufacturer's cost plus one-half the difference between the manufacturer's cost and the market value, as long as it does not exceed twice the cost basis.

This research and experimentation tax credit expired in June 1992 when an extension contained in H.R. 11, the Enterprise Zone Tax Act, was vetoed by former President Bush. The Omnibus Budget Reconciliation Act, P.L. 103-66, reinstated the credit through July 1995 and made it retroactive to the date of its previous expiration. The credit again expired. However, P.L. 104-188, the Small Business Job Protection Act, reinstated the tax credit for application between July 1, 1996 and May 31, 1997. The Taxpayer Relief Act of 1997, P.L. 105-34, extended the credit for 13 months from June 1, 1997 through June 30, 1998. The tax credit expired once again but was reinstated through June 30, 1999, by P.L. 105-277. Several bills also were introduced that would have permitted the research tax credit to be applied to support for certain collaborative research consortia. The 106th Congress once again extended the credit. Title V of P.L. 106-170 reinstated the research and experimentation tax credit through June 30, 2004 and increased the credit rate applicable under the alternative incremental research credit by one percentage point per step. P.L. 108-311 extended the research credit through December 31, 2005 while in the 1 09th Congress, P.L. 109- 432 extended the credit through the end of 2007.[2]

Amendments to the patent and trademark laws contained in P.L. 96-5 17 also were designed to foster interaction between academia and the business community. This law provides, in part, for title to inventions made by contractors receiving federal R&D funds to be vested in the contractor if it is a university, not-for-profit institution, or a small business. Certain rights to the patent are reserved for the government and these organizations are required to commercialize within a predetermined and agreed upon time frame. Providing universities with patent title is expected to encourage licensing to industry where the technology can be manufactured or utilized, thereby creating a financial return to the academic institution. University patent applications and licensing have increased since this law was enacted. (For more discussion on this topic see CRS Report RL32076, *The Bayh-Dole Act: Selected Issues in Patent Policy and the Commercialization of Technology*, by Wendy H. Schacht; CRS Report RL30320, *Patent Ownership and Federal Research and Development (R&D): A Discussion on the Bayh-Dole Act and the Stevenson-Wydler Act*, by Wendy H. Schacht; and CRS Report 98-862, *R&D Partnerships and Intellectual Property: Implications for U.S. Policy*, by Wendy H. Schacht.)

Many cooperative industry-industry or industry-university programs are supported and/or organized by the federal departments and agencies. These include, but are not limited to, the National Science Foundation's Engineering Research Centers, the approximately 40 Industry-University Cooperative Research Programs, and the more recent Science and Technology Centers. A program to match small businesses interested in joint manufacturing technology efforts has been created in the Department of Commerce.

While most legislative activities are intended to facilitate technological advance across industries, there have been several recent efforts to provide direct assistance for cooperative ventures in a particular industry. These initiatives are based, in part, on national defense and economic security concerns over specific technologies that are, or are perceived as, potentially critical to a wide range of businesses. Among the joint ventures, funded primarily by the Department of Defense, have been SEMATECH (a joint private sector semiconductor manufacturing research effort which is now privately financed), the National Center for Manufacturing Sciences, and the steel initiative. In addition, DOD supported the Software Engineering Institute and the Department of Energy assisted in the Partnership for a New Generation Vehicle initiative that, among other things, encouraged joint R&D between federal laboratories and private firms leading to commercialization.

Cooperation between industry and the federal R&D enterprise is another facet of the effort to increase industrial competitiveness through joint ventures. The federal government will spend an estimated $133.7 billion for research and development in FY2006 to meet the mission requirements of the federal departments and agencies. This has led to many technologies and techniques, as well as to the generation of knowledge and skills, which may have applications beyond their original intent. To foster their development and commercialization in the industrial community, various laws have established institutions and mechanisms to facilitate the movement of ideas and technologies between the public and private sectors.

The Stevenson-Wydler Technology Innovation Act (P.L. 96-480), as amended by the Federal Technology Transfer Act (P.L. 99-502) and the Department of Defense FY1990 Authorizations (P.L. 10 1-189), provides, in part, a legislative mandate for technology transfer from the federal government to the private sector, establishes a series of offices in the agencies and/or laboratories to administer transfer efforts, provides incentives for federal laboratory personnel to actively engage in technology transfer, and creates new contractual means for industry to work with the laboratories including cooperative research and development agreements (CRADAs). P.L. 104-113, the National Technology Transfer and Advancement Act, attempts to clarify existing policy with respect to the dispensation of intellectual property under a CRADA by amending the Stevenson-Wydler Act. P.L. 106-404, the Technology Transfer Commercialization Act, makes changes in current practices concerning patents held by the government to make it easier for federal agencies to license such inventions to the private sector for commercialization. (For additional information see CRS Report RL3 3527, *Technology Transfer: Use of Federally Funded Research and Development*, by Wendy H. Schacht.)

The CREATE Act, P.L. 108-453, makes changes in the patent laws to promote cooperative research and development among universities, government, and the private sector. The bill amend section 103(c) of title 25, United States Code, such that certain actions between researchers under a joint research agreement will not preclude patentability. (For more detail see CRS Report RS2 1882, *Collaborative R&D and the Cooperative Research and Technology Enhancement (CREATE) Act*, by Wendy H. Schacht.)

The Omnibus Trade and Competitiveness Act (P.L. 100-418) established a program of regional Centers for the Transfer of Manufacturing Technology (now part of the Manufacturing Extension Partnership effort) to facilitate the movement to the private sector of knowledge and technologies developed under the aegis of the National Institute of Standards and Technology. (For more discussion, see CRS Report 97-104, *Manufacturing Extension Partnership Program*, by Wendy H.

Schacht.) In addition, the law required that NIST provide technical assistance to state technology extension programs in an effort to improve private sector access to federal technology. (For additional Information, see CRS Report RL33528, *Industrial Competitiveness and Technological Advancement: Debate over Government Policy*, by Wendy H. Schacht.) Government-industry collaboration is further facilitated by a provision of the FY1991 National Defense Authorization Act (P.L. 101- 510) that amends Stevenson-Wydler to allow government agencies and laboratories to develop partnership intermediary programs to augment the transfer of laboratory technology to the small business community.

A pilot activity under the Small Business Development Act of 1992, the Small Business Technology Transfer program, facilitates cooperative work between small companies and federal labs leading to the commercialization of new technology. Scheduled to sunset in FY1996, the program was extended for one year until P.L. 105-135 reauthorized it through FY2001. Subsequently, P.L. 107-50 extended the STTR activity through FY2009, increased the set-aside used to fund the program to 0.3% (beginning in FY2004), and expanded the amount of money available for individual Phase II grants to $750,000. (See CRS Report 96-402, *Small Business Innovation Research Program*, by Wendy H. Schacht.)

ISSUES

It is not yet known whether federal support of cooperative ventures signals a long-term commitment to the development of technology. The former Clinton Administration set out a policy to actively promote joint R&D activities utilizing both direct and indirect federal support for expanded cooperative work leading to commercialization. However, given current concerns over the federal budget, it is unlikely that large sums of government money will be forthcoming for such efforts in the future. Yet, other actions may reflect federal interest in the process of technological advancement. The use of the extensive government R&D system, with its expensive state-of-the-art facilities, can provide both academia and industry with resources that may be beyond their financial ability. And despite the often short-term focus of budget decisions, federal funds and non-monetary contributions to cooperative ventures may be leveraged by contributions from state and local agencies and the private sector.

If the proliferation of programs is any indication, state and local jurisdictions have been in the forefront of cooperative endeavors. Many state and local economic development activities focus on increasing innovation and the use of technology in the private sector. Instead of competing for companies to relocate, many of these jurisdictions now see additional benefits accruing from the creation of new firms and the modernization of existing ones through the application of new technology. Various states and localities are attempting to foster an entrepreneurial climate by undertaking the development and support of a variety of programs to assist existing high technology businesses, to promote the establishment of new companies, and to facilitate the use of new technologies and processes in traditional industries. While these efforts vary by state and locality, many of them include industry-university-government cooperation. Several congressional proposals for increasing cooperative ventures built upon existing state and local activities in these areas. (For additional discussion, see CRS Report 96-958 SPR, Technology Development: Federal-State Issues (out of print; contact the author, Wendy H. Schacht, for copies, 202-707-7066) and CRS Report 98-859, State Technology Development Strategies: The Role of High Tech Clusters, by Wendy H. Schacht.)

Proponents of cooperative work argue that certain benefits are associated with joint ventures. The increased popularity of this concept, and expanding federal support for this approach, however, might suggest some questions be raised to assess whether cooperative ventures are meeting expectations. Are there drawbacks to this effort in general and in specific instances? Are cooperative projects addressing the problems associated with the competitiveness of U.S. industry? Are they moving technology development in the right direction?

It might be expected that an increasing number of industries and/or companies will come to the federal government for assistance in supporting cooperative R&D activities. Despite opposition by some to what has been described as "picking winners or losers," various sectors of the government have chosen to provide funding for cooperative ventures in specific industries while requiring that the private sector generate matching funds. At the same time, there are programs and policies that attempt to facilitate cooperative efforts across industry in general. Decisions might need to be made whether one approach is better than the other, or if both should continue.

If part of government policy is to respond to individual industry requests for assistance, Congress may opt to consider developing procedures to select between industries and/or companies competing for limited federal funds. Can, and should, federal guidelines be established? In addition, is it possible to determine at this time what type of cooperative ventures are the most effective and efficient? Is there, in fact, one best model or should each venture be tailored to the specific situation? And **finally, what are the implications of these decisions for policymaking in Congress?**

Development

As noted above, innovation is a dynamic process that can involve idea origination, research, development, commercialization, and diffusion throughout the economy. However, it is not a linear process and an innovation may occur without developing through these steps. In fact, most innovations are actually incremental changes in existing goods and services in response to unmet market needs. The most crucial factor is the availability or use of the technology or technique in the marketplace.

In the recent past, the commercialization and diffusion of products and processes often stood out as significant problems in terms of the ability of U.S. industries to compete. Firms in several other countries, particularly Japan and the East Asian newly industrializing countries, have been successful in commercializing the results of R&D. In various instances, this was research initially performed in the United States, as evidenced by the VCR and semiconductor chips. Basic research and the pursuit of science are done successfully in the United States as indicated, in part, by the number of Nobel prizes awarded to Americans. However, excellence in science does not necessarily assure leadership in world markets. It has been noted that the United States was the world's premiere economic power in the 1920s when this nation was far from being in the forefront of science. Instead, market leadership is significantly affected by the development and application of technology to make the goods and services the consumers want to purchase.

Thus, questions may be raised as to whether programs and policies encouraging increased cooperative research, without concomitant efforts to facilitate the development and commercialization of technologies and techniques, can be effective mechanisms to increase the competitiveness of American industry. Do we need to know more about how to encourage the application of the research resulting from joint ventures in the manufacture of products and processes and in the delivery of services? Do these cooperative activities include mechanisms to facilitate the effective and timely transfer of the results back to the companies where they can be developed into goods for the marketplace? Since the major portion of the costs associated with bringing out a new product occur at the development and marketing stages, not in the research phase, should there be additional government incentives to encourage companies to spend funds for commercialization in addition to research?

Manufacturing

It is in the manufacturing arena where American companies appear to be the most vulnerable to foreign competition. Process technologies (those used in manufacturing) can significantly lower the costs of production and increase the quality of goods and services. In Global Competition, the President's Commission on Industrial Competitiveness (under former President Reagan) concluded that "... competitive success in many industries today is as much a matter of mastering the most advanced manufacturing processes as it is in pioneering new products."

The costs associated with the development and purchase of new manufacturing equipment are high. This is particularly true for the 350,000 small companies which make up a major segment of the manufacturing community. Several of the cooperative efforts supported by the federal government address these manufacturing concerns. The Manufacturing Technology program of the Department of Defense, the advanced manufacturing initiatives in the Department of Energy, and the Manufacturing Extension Centers operated by the National Institute of Standards and Technology, although all different, are examples of government activities devoted to facilitating the development of new manufacturing techniques and their use in industry.

Considering the importance of manufacturing, the existing cooperative programs may not be sufficient to increase the competitiveness of American industry. Are there more effective types of joint ventures? Cooperative efforts, where resources could be pooled and the equipment shared, may be one way to improve the manufacturing capability of U.S. firms, large or small. Will joint manufacturing prove to be a viable option? Should existing cooperative manufacturing programs in certain agencies be expanded or should new efforts in other departments be developed? Should one government agency have the lead in policy determinations; if so, which federal department?

Defense vs. Civilian Support

Many of the industries interested in cooperative ventures with federal financial support have approached the Department of Defense and, to a lesser extent, the Department of Energy's Defense Programs because these agencies have the greatest amount of available resources and/or funding. They also tend to have the expertise to operate large-scale programs and maintain close ties with certain industrial sectors which could be encouraged to increase cooperation. In addition, both DOD and DOE have a vested interest in the availability of certain technologies which could be provided by a healthy domestic commercial market. However, questions remain whether sponsorship of certain cooperative ventures by DOD and the Department of Energy's defense-related programs will lead to increased commercialization in the civilian marketplace.

Critics argue that defense spending is not an effective mechanism to increase industry's ability to innovate and develop new technologies. Much of the research and development in the defense arena may be too specialized, overdesigned, and/or too costly to have value for commercial markets. The R&D also tends to concentrate on weapon systems and other defense hardware rather than on process technologies that are often necessary to improve manufacturing productivity. One reason cited for the competitive problems of the machine tool industry was its focus on defense needs rather than on the commercial market which is larger in the aggregate.

On the other hand, the U.S. commitment to military R&D has contributed to a favorable balance of trade in the defense and aerospace industries. In the SEMATECH effort, the purpose of DOD support was to facilitate the *commercial* development of technologies with critical defense applications. The companies involved in SEMATECH were experienced semiconductor manufacturers and were knowledgeable about the markets' needs and operations. Thus, although the initial work performed by this semiconductor consortium may have been partially funded by the Defense Advanced Research Project Agency, it was designed to result in new products and processes in the civilian marketplace where both defense and commercial demand can be met. SEMATECH now operates without direct federal financing.

The issue of cooperative work between the Defense Department and the private sector leading to commercial technologies was addressed in the former Technology Reinvestment Project and was part of the more recent Dual-Use Partnership Project. The Department of Energy has been expanding cooperative R&D activities in Defense Program laboratories in conjunction with an increase in all DOE collaborative efforts with industry. Recent significant decreases in the technology transfer budgets may impeded this effort, but several DOE defense laboratories are actively pursuing joint ventures with industry. (See CRS Report 98-8 1, *Cooperative Research and Development Agreements and Semiconductor Technology: Issues Involving the "DOE-Intel CRADA ",* by Wendy H. Schacht and Glenn J. McLoughlin.)

Access by Foreign Firms

With worldwide communications systems, it is virtually impossible to prevent the flow of scientific and technical information. What is critical to competitiveness is the speed at which this knowledge is used to make products, processes, and services for the marketplace. However, it appears that many foreign firms are willing and able to take the results of research performed both in the United States and their own countries and rapidly make high quality commercial goods. Many of these companies are purchasing American businesses or establishing U.S. subsidiaries to access American expertise. With the increased activity in research consortia, particularly those with federal support, questions might be asked as to whether or not foreign companies should or could be barred from access to the results. A larger issue is how to define an "American company." Is it determined by majority ownership, manufacturing, location, value added to the U.S. economy, or by some other definition? In addition, since technology is most effectively transferred by person-to-person interaction, would cooperative activities between American industry and foreign firms produce an outflow of information which could be used to increase competitive pressures?

Direct vs. Indirect Support

Government efforts to facilitate cooperative ventures have included both indirect supports and direct federal funding. Indirect measures include such things as tax policies, intellectual property rights, and antitrust laws that create incentives for the private sector. Other initiatives include government financing (on a cost shared basis) of joint efforts such as the Advanced Technology Program and Manufacturing Extension Partnerships. In the past, participants in the legislative process generally did not make definite (or exclusionary) choices between these two approaches. However, these activities were revisited in the 104th Congress given apparent Republican preferences for the funding of basic research and not technology development. For example, efforts to eliminate the Advanced Technology Program, funding for flat panel displays, and agricultural extension reflected concern over the role of government in developing commercial technologies and generally resulted in reductions of direct federal financing for such public-private partnerships. Issues were again raised in the subsequent Congresses although no relevant, on-going program was terminated. As the 1 10th Congress makes its budget decisions, the future of cooperative R&D may be expected to be explored further. (For more information, see CRS Report 95-50, *The Federal Role in Technology Development*, by Wendy H. Schacht.)

109ᵀᴴ CONGRESS LEGISLATION

P.L. 109-108 (H.R. 2862)

Makes appropriations for science and the Departments of State, Justice, and Commerce. As passed by the House, the bill would provide $106 million for the Manufacturing Extension Partnership and no financing for the Advanced Technology Program. The version of the legislation reported to the Senate from the Committee on Appropriations would fund MEP at $106 million and provide $140 for ATP. The final appropriations legislation financed MEP at $104.6 million and ATP at $79 million (after mandated rescissions). Introduced June 10, 2005; referred to the House Committee on Appropriations. Passed the House, amended, on June 16, 2005. Received in the Senate on June 16, 2005; referred to the Senate Committee on Appropriations. Reported to the Senate, with an amendment in the nature of a substitute, on June 23, 2005. Passed the Senate, amended, on September 15, 2005. Conference report filed November 7, 2005. House agreed to conference report on November 9, 2005; Senate agreed on November 16, 2005. Signed into law by the President on November 22, 2005.

P.L. 109-432 (H.R. 6111)

Amends the Internal Revenue Code of 1986 to extend the research and experimentation tax credit through the end of 2007, among other things. Introduced September 19, 2006; referred to the House Committee on Ways and Means. Passed the House, amended on December 5, 2006. Passed the Senate, amended, by unanimous consent, on December 7, 2006. The Senate agreed to the House amendment by unanimous consent on December 9, 2006. Signed into law by the President on December 20, 2006.

H.R. 250 (Ehlers)/S. 2134 (Smith)

Manufacturing Technology Competitiveness Act. Creates an interagency committee to coordinate federal manufacturing R&D. Establishes and authorizes funding for a pilot collaborative manufacturing research grants program to promote the development of new manufacturing technologies through cooperative applied research among the private sector, academia, states, and other non-profit institutions. Mandates and authorizes financing for a manufacturing fellowship program. Creates and authorizes support for a manufacturing extension center competitive grants program to focus on new or emerging manufacturing technologies. Authorizes funding for the Manufacturing Extension Partnership, among other things. Introduced January 6, 2005; referred to the Committee on Science. Reported to the House, amended, May 23, 2005. Passed House on September 21, 2005. Received in Senate and referred to the Senate Committee on Commerce, Science, and Transportation on September 22, 2005. S. 2134 introduced December 16, 2005; referred to the Senate Committee on Commerce, Science, and Transportation.

H.R. 3331 (Miller, B.)

Creates and authorizes funding for a grant program in the National Science Foundation to assist universities in promoting the application of new inventions developed within their institutions. Introduced July 27, 2005; referred to the House Committee on Science.

H.R. 4845 (Goodlatte)

Innovation and Competitiveness Act. Makes permanent the research and experimentation tax credit, among other things. Introduced March 2, 2006; referred to the House Committees on the Judiciary, Ways and Means, science, Education and the Workforce, and Energy and Commerce.

H.R. 5672 (Wolf)

FY2007 Science, State, Justice, Commerce, and Related Agencies Appropriations Act. Funds the Manufacturing Extension Partnership at $92 million but provides no support for the Advanced Technology Program, among other things. Introduced on June 22, 2006; referred to the House Committee on Appropriations. Reported to the House on June 22, 2006 and passed the House on June 29, 2006. Reported from the Senate Committee on Appropriations, with an amendment in the nature of a substitute, on July 13, 2006.

S. 1581 (Bingaman)

Provides financing and other assistance (including tax incentives for private sector investments) for the development of science parks, among other things. Introduced July 29, 2005; referred to the Senate Committee on Finance.

S. 2109 (Ensign)/H.R. 4654 (Schiff)/S. 2390 (Ensign)

National Innovation Act. Establishes a President's Council on Innovation and provides innovation acceleration grants. Promotes innovation through regional economic development and makes permanent the research and experimentation tax credit, among other things. S. 2109 introduced December 15, 2005; referred to the Senate Committee on Finance. H.R. 4654 introduced January 3, 2006; referred to the House Committees on Science, Energy and Commerce, Ways and Means, Armed Services, Judiciary, Transportation and Infrastructure, and Financial Services. S. 2390 introduced March 8, 2006; referred to the Senate Committee on Commerce, Science, and Transportation.

S. 2198 (Domenici)

Protecting America's Competitive Edge Through Education Act. Among other things creates mechanisms to develop and fund Science Parks. Introduced January 26, 2006; referred to the Senate Committee on Health, Education, Labor, and Pensions. Hearings held February 28, 2006 and March 1, 2006.

S. 2199 (Domenici)

Protecting America's Competitive Edge Through Tax Incentives Act. Expands and makes permanent the research and development tax credit, among other things. Introduced January 26, 2006; referred to the Senate Committee on Finance.

S. 2720 (Baucus)

Research Competitiveness Act of 2006. Simplifies the research tax credit and makes it permanent. Allows for tax exempt financing of research park facilities, among other things. Introduced May 4, 2006; referred to the Senate Committee on Finance.

S. 2802 (Ensign)

American Innovation and Competitiveness Act of 2006. Among other things, establishes the President's Council on Innovation and Competitiveness, provides innovation acceleration grants, and facilitates regional economic development. Introduced May 15, 2006; referred to the Senate Committee on Commerce, Science, and Transportation. Reported, with amendments, on July 18, 2006.

LEGISLATION IN THE 110TH CONGRESS

H.R. 85 (Biggert)

Energy Technology Transfer Act. Creates a program of grants to non-profit institutions, state and local governments, cooperative extension services, or universities to transfer energy efficient methods and technologies. Introduced January 4, 2007; referred to the House Committee on Science and Technology.

H.R. 255 (Ehlers)

Manufacturing Technology Competitiveness Act of 2007. Creates an interagency committee to coordinate federal manufacturing R&D. Establishes and authorizes funding for a pilot collaborative manufacturing research grants program to promote the development of new manufacturing technologies through cooperative applied research among the private sector, academia, states, and other non-profit institutions. Mandates and authorizes financing for a manufacturing fellowship program. Creates and authorizes support for a manufacturing extension center competitive grants program to focus on new or emerging manufacturing technologies. Authorizes funding for the Manufacturing Extension Partnership, among other things. Introduced January 5, 2007; referred to the House Committee on Science and Technology.

H.R. 363 (Gordon)

Sowing the Seeds Through Science and Engineering Research Act. Authorizes a Presidential Innovation Award, among other things. Introduced January 10, 2007; referred to the House Committee on Science and Technology. Reported from the Committee, amended, on March 8, 2007. Passed the House on April 24, 2007. Received in the Senate and referred to the Senate Committee on Health, Education, Labor and Welfare on April 25, 2007.

H.R. 1712 (Johnson, E.B.)

The Research and Development Tax Credit Act of 2007. Makes the research tax credit permanent and allows for the issuance of tax exempt facility bonds for research park facilities used for research and experimentation, among other things. Introduced March 27, 2007; referred to the House Committee on Ways and Means.

H.R. 1868 (Wu)

Technology Innovation and Manufacturing Stimulation Act of 2007. Authorizes funding for the National Institute of Standards and Technology through 2010 and creates several new manufacturing R&D programs in that organization. Introduced April 17, 2007; referred to the House Committee on Science and Technology.

S. 41 (Baucus)

Research Competitiveness Act of 2007. Amends the Internal Revenue Code to make the research and experimentation tax credit permanent. Among other things, this bill would allow the issuance of tax exempt facility bonds for research park facilities used for research and experimentation. Introduced January 4, 2007; referred to the Senate Committee on Finance.

S. 69 (Kohl)

Authorizes appropriations for the Manufacturing Extension Partnership through 2012, among other things. Introduced January 4, 2007; referred to the Senate Committee on the Judiciary. Discharged from the Senate Committee on the Judiciary by unanimous consent on January 22, 2007 and referred to the Senate Committee on Commerce, Science, and Transportation the same day.

S. 592 (Collins)

GoMe Act. Extends the research tax credit through 2012, among other things. Introduced February 14, 2007; referred to the Senate Committee on Finance.

S. 761 (Reid)

America Creating Opportunities to Meaningfully Promote Excellence in Technology, Education, and Science Act. Mandates a National Science and Technology Summit to access the state of U.S. science and technology. Requires a study on barriers to innovation and creates a National Innovation Medal and a President's Council on Innovation and Competitiveness. Requires that federal agencies establish an Innovation Acceleration Research Program to facilitate innovation, among other things. Introduced March 5, 2007; placed on Senate

Legislative Calendar under General Orders March 6, 2007. Passed the Senate, amended, on April 25, 2007.

S. 833 (Coleman)

COMPETE Act of 2007. Makes the research and experimentation tax credit permanent, among other things. Introduced March 9, 2007; referred to the Senate Committee on Finance.

REFERENCES

[1] National Science Board, *Science and Engineering Indicators 2006*, available at [http://www.nsf.gov/statistics/seind06/c4/c4s5.htm#fn3 8].

Chapter 3

THE BAYH-DOLE ACT: SELECTED ISSUES IN PATENT POLICY AND THE COMMERCIALIZATION OF TECHNOLOGY[*]

Wendy H. Schacht

ABSTRACT

Congressional interest in facilitating U.S. technological innovation led to the passage of P.L. 96-5 17, Amendments to the Patent and Trademark Act (commonly referred to as the Bayh-Dole Act after its two main sponsors). The act grants patent rights to inventions arising out of government-sponsored research and development (R&D) to certain types of entities with the expressed purpose of encouraging the commercialization of new technologies through cooperative ventures between and among the research community, small business, and industry.

Patents provide an economic incentive for companies to pursue further development and commercialization. Studies indicate that research funding accounts for approximately one-quarter of the costs associated with bringing a new product to market. Patent ownership is seen as a way to encourage the additional, and often substantial investment necessary for generating new goods and services. In an academic setting, the possession of title to inventions is expected to provide motivation for the university to license

---

[*] Excerpted from CRS Report RL32076, dated April 3, 2008.

the technology to the private sector for commercialization in expectation of royalty payments.

The Bayh-Dole Act has been seen as particularly successful in meeting its objectives. However, while the legislation provides a general framework to promote expanded utilization of the results of federally funded research and development, questions are being raised as to the adequacy of current arrangements. Most agree that closer cooperation among industry, government, and academia can augment funding sources (both in the private and public sectors), increase technology transfer, stimulate more innovation (beyond invention), lead to new products and processes, and expand markets. However, others point out that collaboration may provide an increased opportunity for conflict of interest, redirection of research, less openness in sharing of scientific discovery, and a greater emphasis on applied rather than basic research. Additional concerns have been expressed, particularly in relation to the pharmaceutical and biotechnology industries, that the government and the public are not receiving benefits commensurate with the federal contribution to the initial research and development.

Actual experience and cited studies suggest that companies which do not control the results of their investments — either through ownership of patent title, exclusive license, or pricing decisions — tend to be less likely to engage in related R&D. The importance of control over intellectual property is reinforced by the positive effect P.L. 96-517 has had on the emergence of new technologies and techniques generated by U.S. companies.

INTRODUCTION

Congressional interest in facilitating U.S. technological innovation led to the passage of P.L. 96-5 17, Amendments to the Patent and Trademark Act, commonly referred to as the "Bayh-Dole Act" after its two main sponsors former Senators Robert Dole and Birch Bayh. Under this 1980 law, as amended, title to inventions made with government support is provided to the contractor if that contractor is a small business, a university, or other non-profit institution. The legislation is intended to use patent ownership as an incentive for private sector development and commercialization of federally funded research and development (R&D). As a response to congressional efforts to create a unified government patent policy pertaining to inventions made with federal support, the Bayh-Dole Act promotes cooperative activities among academia, small business, and industry leading to new products and processes for the marketplace.

This paper discusses the rationale behind the passage of P.L. 96-517, its provisions, and implementation of the law. Observers generally agree that the Bayh-Dole Act has successfully met its objectives. However, some experts argue that the issues associated with the law's patent policies should be revisited given the current R&D environment. Much of the renewed interest is a result of the legislation's effect on the biotechnology and pharmaceutical industries where critics assert that the private sector is receiving benefits to the detriment of the public interest. Other analysts, particularly in the defense arena, maintain that the existing rights maintained by the government are too restrictive and prevent industry from meeting national needs. Many of these issues and concerns are similar, if not identical to those addressed during the 15 to 20 years of deliberation prior to enactment of the law. These too will be explored to provide a context for current discussions.

AN HISTORICAL PERSPECTIVE

The Rationale

In the late 1970s, the United States Congress was involved in a series of legislative debates over ways to promote private sector development and utilization of federally funded research and development. This was soon followed by expanded congressional interest in additional means to foster technological advancement and commercialization in industry. During the 1980s and 1990s, various initiatives resulted in laws designed to encourage increased innovation-related activities in the business community and to remove barriers to technology development, thereby permitting market forces to operate.[1] Laws promoting cooperative R&D and/or joint ventures involving the federal government, industry, and academia have been a cornerstone of the majority of these efforts and include legislation that created a system to transfer technology from federal laboratories to the private sector; implemented tax incentives for collaborative work; instituted direct and indirect government support for increased R&D; and changed government patent policy to provide an economic inducement for commercialization of federally funded technology, the subject of this report.

P.L. 96-5 17, the Bayh-Dole Act, was one of the first of these initiatives. Prior to 1980, only 5% of government owned patents were ever used in the private sector although a portion of the intellectual property portfolio had potential for further development, application, and marketing. The Bayh-Dole Act was constructed, in part, to address the low utilization rate of these federal patents. The House report to accompany H.R. 6933 (the House counterpart to the Senate bill that eventually became the Bayh-Dole Act) noted that, at the time the bill was considered, 26 different agency policies existed regarding the use of the results of federally funded

R&D. Generally the government retained title to inventions made with government support whether the research was performed in federal laboratories, in universities, or by individual companies. Licenses to use government patents were then negotiated with firms either on a non-exclusive basis (meaning additional companies could use the technology) or, more rarely, for the exclusive use by one manufacturer. However, it was widely argued that without title (or at least an exclusive license) to an invention and the protection it conveys, a company would not invest the additional, and often substantial time and money necessary to commercialize a product or process for the marketplace.

In 1980, the federal expenditure for research and development totaled $55.5 billion (in constant 2000 dollars).[2] The money typically was used to support research and development to meet the mission requirements of the federal departments and agencies (e.g., defense, public health, environmental quality) or to finance work in areas where there was an identified need for research, primarily basic research, not being performed in the private sector. While the government's investment led to many new inventions that have profoundly influenced our society, many in Congress were of the opinion that additional applications could be pursued by the private sector if provided the proper incentives.

The intent of the new law was to replace this situation with a "single, uniform national policy designed to cut down on bureaucracy and encourage private industry to utilize government financed inventions through the commitment of the risk capital necessary to develop such inventions to the point of commercial application."[3] Expanded technology commercialization was to be accomplished by employing the patent system to augment collaboration between universities (as well as other nonprofit institutions) and the business community to ensure that inventions are brought to market. The Bayh-Dole Act also provides for the increased participation of small firms in the national R&D enterprise under the assumption that these companies tend to be more innovative than larger companies.

The Patent System: A Brief Overview

The patent system was created to promote invention and innovation. Article I, Section 8, Clause 8 of the U.S. Constitution states: "The Congress Shall Have Power ...To promote the Progress of Science and useful Arts, by securing for limited Times to Authors and Inventors the exclusive Right to their respective Writings and Discoveries..." Patents are widely believed to encourage innovation by simultaneously protecting the inventor and fostering competition. They provide the inventor with a right to exclude others, temporarily, from use of the invention without compensation. Patents give the owner an exclusive right for 20 years (from date of filing) to further develop the idea, commercialize a product or process, and potentially realize a return on the initial investment. Concurrently, the process of obtaining a patent places the concept in the public arena. As a disclosure system, the patent can, and often does, stimulate other firms or individuals to invent "around" existing patents to provide for parallel technical developments or meet similar market needs.[4]

Not everyone agrees that the patent system facilitates innovation. Critics argue that patents provide a monopoly which induces additional social costs. Others assert that the patent system is unnecessary due to market forces that already suffice to create an optimal level of innovation. The desire to obtain a lead time advantage over competitors, as well as the recognition that technologically backward firms lose out to their rivals, may well provide sufficient inducement to invent without the need for further incentives.[5] Some commentators believe that the patent system encourages industry concentration and presents a barrier to entry in some markets and that cross licensing between companies can result in exploitation of other markets.[6] Still other observers believe that the patent system too frequently attracts speculators who prefer to acquire and enforce patents rather than engage in socially productive activity.[7]

The importance of patents varies among industrial sectors. Patents are perceived as critical in the drug and chemical industries in part because of the ease of replicating the finished product. While it is expensive, complicated, and time consuming to duplicate an airplane, it is relatively simple to chemically analyze a pill and reproduce it.[8] Studies have found that in many other industries the protection offered by patents is diminished by the ability to invent around the patent and limited by the disclosure of vital information in the patent itself.[9] In the aircraft and semiconductor industries, patents have not been the most successful mechanism for capturing the benefits of investments. Instead, lead time and the strength of the learning curve were determined to be more important.[10] Later studies bear this out; secrecy and lead time were deemed to have greater effect than patents in the semiconductor and related equipment industry, as well as the aerospace and machine tool industries, among others.[11]

Patents can provide an economic incentive for companies to pursue further development and commercialization. Studies indicate that research funding accounts for approximately one-quarter of the costs associated with bringing a new product to market. According to *The Economist*, "A dollar's worth of academic invention or discover requires upwards of $10,000 of private capital to bring [it] to market."[12] Patent ownership is seen as a way to encourage the additional, and often substantial investment necessary for new goods and services, particularly in the case of small business. In an academic setting, the possession of title to inventions is expected to provide motivation for the university to license the technology to the private sector for commercialization in anticipation of royalty payments.

University-Industry Cooperation

Changes to the patent laws embodied in the Bayh-Dole Act had as an objective the facilitation of collaborative ventures between and among academia, industry, and government. In 1980, universities performed 14% of the R&D undertaken in the United States (similar to today); much of this the fundamental research basic to technological advance.[13] The work is accomplished as part of the education process and provides training for scientists, engineers, and managers subsequently employed by the private sector.

Universities, however, generally do not have the means of production necessary to take the results of research and generate marketable products. Such activities are carried out by industry. Thus, the emphasis in the Bayh-Dole Act on the promotion of cooperative efforts between academia and the business community. By providing universities with intellectual property ownership with which to pursue and structure collaborative ventures, the legislation is intended to encourage the two sectors to work together to generate new goods, processes, and services for the marketplace. Such joint work allows for shared costs, shared risks, shared facilities, and shared expertise.

Prior to World War II, industry was the primary source of funding for basic research in universities. This financial support helped shape priorities and build relationships. However, after the war, the federal government supplanted the private sector as the major financial contributor and became the principal determinant of the type and direction of the research performed in academic institutions. This situation oftentimes resulted in a disconnect between the university and industrial communities. Because the private sector and not the government typically is involved in commercialization, the difficulties in moving an idea from the research stage to a marketable product or process appeared to have been compounded. Thus, efforts to encourage increased collaboration between and among the sectors through the Bayh-Dole Act were expected to augment the contribution of both parties to technological advancement.

Small Business

Special consideration concerning patent title was given to small businesses in part because of the role these companies were seen as playing in the generation of new jobs and in technological advancement. Early research supported by several federal agencies concluded that small, high technology companies are the source of significant innovation. An often cited 1982 study financed by the Small Business Administration determined that small firms were 2.4 times as innovative per employees as large companies.[14] Similar work performed at the time the legislation was being considered found that firms of less than 1,000 employees were responsible for more major innovations than large firms in the years 1953-1966 and for an equal number from 1967-1973.[15] A study of national and regional data by the Federal Reserve Bank of Chicago concluded that "small firms — those with 20 or fewer employees — create a larger proportion of new jobs than their share of employment in the economy and continue to create jobs even during recession."[16]

However, certain caveats need to be stated particularly within the context of small business, innovation, and technology development. Over the years, experts have argued that the contribution of small firms to the economy is overstated. Marc Levinson, writing in Dun's Business Month during the 1980s, maintained that small companies tended to produce fewer goods than larger ones because they are less capital intensive and, on the whole, add less to the gross national product because they offer lower salaries and often do not provide health insurance or pension plans.[17] Professors Zoltan Acs and David Audretsch argued that the relationship between company size and innovation capacity varied by industry.[18] Others maintained that there is no conclusive evidence that firm size affects the "success" of R&D.[19]

An important factor affecting the ability of small companies to effect technological advance appears to be the relationship between these firms and large corporations, a concept that is reflected in the provisions of the Bayh-Dole Act:

> the corporate contribution and that of the innovative entrepreneur are characteristically very different from one another and characteristically play complementary roles. Moreover, the contribution of the two together is superadditive, that is, the combined result is greater than the sum of their individual contributions.[20]

A study by the National Academy of Engineering concluded that "small high-tech companies play a critical and diverse role in creating new products and services, in developing new industries, and in driving technological change and growth in the U.S. economy."[21] The reasons for this include the ability of these firms to rapidly develop markets, generate new goods and services and offer product diversity. Small businesses tend to be willing to take those technological risks that are not pursued by large firms and may be in a position to quickly exploit market opportunities. In specific cases, experts note, "an innovative disadvantage of large firms is an innovative advantage for small firms, and vice versa, which can make collaboration between two firms of different size desirable for both parties."[22]

BAYH-DOLE AND RELATED LAW

Provisions

In enacting P.L. 96-517, the Congress accepted the proposition that vesting title to the contractor will encourage commercialization and that this should be used to support innovation in certain identified sectors. The law states:

> It is the policy and objective of the Congress to use the patent system to promote the utilization of inventions arising from federally-supported research or development; ...to promote collaboration between commercial concerns and nonprofit organizations, including universities; ...to promote the commercialization and public availability of inventions made in the United States by United States industry and labor; [and] to ensure that the Government obtains sufficient rights in federally-supported inventions to meet the needs of the Government and protect the public against nonuse or unreasonable use of inventions....[23]

Each nonprofit organization (including universities) or small business is permitted to elect (within a reasonable time) to retain title to any "subject invention" made under federally funded R&D; except under "exceptional circumstances when it is determined by the agency that restriction or elimination of the right to retain title to any subject invention will better promote the policy and objectives of this chapter."[24] The institution must commit to commercialization within a predetermined, agreed upon, time frame. As stated in the House report to accompany the bill, "the legislation establishes a *presumption* [emphasis added] that ownership of all patent rights in government funded research will vest in any contractor who is a nonprofit research institution or a small business."[25]

Certain rights are reserved for the government to protect the public's interests. The government retains "a nonexclusive, nontransferable, irrevocable, paid-up license to practice or have practiced for or on behalf of the United States any subject invention throughout the world...." The government also retains "march-in rights" which enable the federal agency to require the contractor (whether he owns the title or has an exclusive license) to "grant a nonexclusive, partially exclusive, or exclusive license in any field of use to a responsible applicant or applicants...." (with due compensation) or to grant a license itself under certain circumstances. The special situation necessary to trigger march-in rights involves a determination that the contractor has not made efforts to commercialize within an agreed upon time frame or that the "action is necessary to alleviate health or safety needs which are not reasonably satisfied by the contractor...."[26]

The government is "authorized" to withhold public disclosure of information for a "reasonable time" until a patent application can be made. Licensing by any contractor retaining title under this act is restricted to companies which will manufacture substantially within the United States. Initially, universities were limited in the time they could grant exclusive licenses for patents derived from government sponsored R&D to large companies (5 of the *then* 17 years of the patent). This restriction, however, was voided by P.L. 98-620, the Trademark Clarification Act of 1984. According to S.Rept. 98-662, extending the time frame for licensing to large firms "is particularly important with technologies such as pharmaceuticals, where long development times and major investments are usually required prior to commercialization."[27]

Most experts continue to argue that patent exclusivity is important for both large and small firms. In a February 1983 memorandum concerning the vesting of title to inventions made under federal funding, then President Ronald Reagan ordered all agencies to treat, as allowable by law, all contractors regardless of size the same as prescribed in P.L. 96-517. This, however, does not have a legislative basis. P.L. 98-620, noted above, further amended Bayh-Dole by loosening the time limitations for both disclosure of an invention to the government agency and for the amount of time provided within which to elect to take title. Nonprofit institutions were subsequently permitted to assign title rights to another organization (e.g., one which markets technology) and government-owned, contractor-operated laboratories (primarily those of the Department of Energy) run by nonprofits were permitted to retain title to inventions made in the facility with the exception of those dedicated to naval nuclear propulsion or weapons development. In addition, the Federal Technology Transfer Act (P.L. 99-502) allows firms regardless of size to be awarded patents generated under a cooperative research and development agreement (CRADA) with a federal laboratory.[28]

Implementation and Results

The Bayh-Dole Act appears to have met its expressed goals of using "the patent system to promote the utilization of inventions arising from federally-supported research or development; ... and to promote collaboration between commercial concerns and nonprofit organizations, including universities...."[29] In one of the earliest studies of the legislation, the General Accounting Office (now the Government Accountability Office, GAO) found agreement among university administrators and small business representatives that P.L. 96-517 had "a significant impact on their research and innovation efforts."[30] While noting it was not correct to generalize about academia from the 25 universities studied, GAO did find that by 1987 all university administrators questioned indicated that the Bayh-Dole Act had "been significant in stimulating business sponsorship of university research, which has grown 74 percent..." from FY1980 to FY1985.[31] According to the National Science Foundation (NSF), industry support for academic research grew faster than any other funding source until FY2002. Industry financing expanded from 3.9% of university R&D in 1980 to 7.2% in 2000, although by FY2006 industry support had dropped to 5.1% of academic R&D. In 1980, federal financing comprised 67.5% of the total academic undertaking; by 2000 federal support declined to 58.2% of university funding, yet increasing to 62.9% in FY2006.[32] It should be noted, however, that the federal government still remains the major source of academic research funding.

The majority of the university personnel involved in the GAO study indicated that the increase in industry support for research at universities was "directly" attributed to the patent changes in P.L. 96-5 17 and P.L. 98-620. Academic faculty interviews conducted by GAO found that "since businesses knew that universities could take title to federally funded inventions, they no longer were concerned that their research efforts could be 'contaminated' by federal funding with the possibility that a federal agency could assert title rights to resulting inventions."[33] All respondents agreed that the removal of licensing restrictions on nonprofit institutions (including universities) by P.L. 9 8-620 was of vital importance in promoting industry-university interaction.[34] This was reinforced by the finding that 9 out of 10 business executives questioned identified the Bayh-Dole Act as an "important factor" in their decisions to fund R&D in academia.[35]

Another GAO study published in May of 1998 reported that agency and university representatives believed the Bayh-Dole Act was meeting its goals as articulated by the Congress and the law had a positive impact on all involved. Academia was "receiving greater benefits from their inventions and were transferring technology better than the government did when it retained title to inventions."[36] In addition, the report states that the increased commercialization of federally funded research that resulted from the implementation of the act positively affected both the federal government and the American people.[37]

Other experts agree. Yale President Richard Levin argued that the purpose of the Bayh-Dole Act is to transition the results of government funded research "into practice for the benefit of humanity..." and that results indicate a "pretty emphatic positive answer that the Bayh-Dole Act has created public benefits" with minimal costs.[38] As stated in an article in *The Economist*, the Bayh-Dole Act is "[p]robably the most inspired piece of legislation to be enacted in America over the past halfcentury...."[39]

One of the major factors in the reported success of the Bayh-Dole Act is the certainty it conveys concerning ownership of intellectual property. The Director of Stanford University's Office of Technology Licensing, Katherine Ku, noted that exclusivity is what motivates firms to invest financial and human resources in technology development.[40] It provides an incentive for universities to take the time and effort to pursue a patent and to license those patents in its portfolio. This has led to a significant increase in academic patenting. In 1980, 390 patents were awarded to universities;[41] by 2005, this number increased to 2,725.[42]

Academia has become a major source of innovation for local and regional economic development. In the latest published survey (FY2006) performed by the Association of University Technology Managers (AUTM), universities identified 697 new products that were marketed that year based on academic R&D. In addition, the survey indicated that during FY2006 more than 553 new companies were created to commercialize university research, 4,963 new licenses/options were granted, with small businesses primarily responsible for the commercialization. Since 1980, more than 5,724 new companies have been established to develop and market academic R&D.[43]

Many of the start-up businesses created from university R&D were associated with just seven schools including the Massachusetts Institute of Technology (MIT), the University of California, California Tech, the University of Minnesota, the Johns Hopkins University, the University of Utah, and the University of Virginia.[44] While only a few universities earn large returns from licensing,[45] studies indicated that licensing by the University of California system generates $91 million in net licensing income annually with Columbia University receiving approximately $80 million and Florida State University $45 million.[46] During FY2004, the Association of University Technology Managers found that $1.4 billion in royalties were generated from 11,414 university licenses.[47]

However, several analysts argue that "Bayh-Dole was only one of a number of important factors behind the rise of university patenting and licensing activity."[48]

In a study of the technology transfer and patenting activities of the University of California, Stanford University, and Columbia University, Professor David Mowery and his colleagues concluded that increased federal funding for basic biomedical research, expanded research in biotechnology, specific court rulings, and government policies augmenting what can be patented all contributed to the rise in academic intellectual property activities. According to their assessment, the Bayh-Dole Act had "little impact on the content of academic research." The pursuit of patenting and licensing at universities has expanded because of changes in biomedical and biotechnology R&D, not because of the act.[49] Later work by Professor Mowery follows this approach, arguing that "the emphasis on the Bayh-Dole Act as a catalyst to these interactions [increased university-industry cooperation and technology transfer] also seems somewhat misplaced, ignoring as it does the long history, extending to at least the earliest decades of the 20th century, of collaboration and knowledge flows between universities and industry in the United States."[50]

Other experts criticize this assessment and point out that the act had the most significant impact on universities that were not actively engaged in patenting prior to its passage.[51] Proponents of this position argue that as a result of the Bayh-Dole Act, in part, "university patenting increased particularly rapidly during the second half of the 1980s and early 1990s."[52] This growth in patenting has been concentrated in "middle-tier" schools, not just the top research universities.[53] Researchers who take this position suggest that the Mowery et al. study focused solely on universities that were previously involved in patenting and licensing and may not have fully considered patent problems that existed before the legislation was implemented. According to critics of the study, the analysts also failed to take into account changes in the venture capital industry that promoted the development of start-up companies to commercialize the results of university R&D.[54]

Other research questions the effect of increased university licensing on U.S. innovation. A study by Bhavan Sampat suggests that while the Bayh-Dole Act augmented patent and licensing by universities, these activities are just "...one of many channels through which universities make economic contributions and in most industries less important contributions than those made by placing scientific and technological information in the public domain."[55] This author's work indicates that "...there is little evidence that increased university patenting and licensing has facilitated increased technology transfer or any meaningful growth in the economic contributions of universities."[56]

However, commentators argue that the provisions of the Bayh-Dole Act provide incentives to take university inventions and develop them into products for the marketplace.[57] University technology generally is in the early stage and not yet ready for commercialization, requiring additional funding and the involvement of faculty to move the idea into a marketable product.[58] While most universities do not receive large amounts of funds as a result of licensing their technologies, it

> is clear from the evidence. ...that faculty involvement in the further development of university technologies is an important element in getting those technologies to market. Mechanisms to ensure such efforts are an important element of commercialization regardless of whether those mechanisms included licensing by universities.[59]

In addition, Professor Scott Shane observes:

> Because universities exploit their inventions primarily through the licensing of technology, and licensing is not equally effective across all technologies.. .the incentive to become more commercially focused led universities to concentrate their patenting in fields in which knowledge is transferred effectively through licensing.[60]

While the effects of the Bayh-Dole Act on the small business sector have not been as extensively studied, the results appear similar. All eight small business owners interviewed by GAO for its 1987 study indicated that the patent changes had a significant beneficial effect on research, development, and innovation in their firms.[61] Perhaps most illustrative of the influence of the Bayh-Dole Act on small business is the biotechnology industry. According to Dr. Bernadine Healy, the former Director of the National Institutes of Health (NIH), P.L. 96-5 17 was responsible for the development and growth of the biotechnology sector.[62] The biotechnology industry primarily is composed of small firms that are developing technologies and techniques derived from R&D funded by NIH. Many of these companies have been established by NIH alumni or university professors previously supported by NIH grants. In Senate testimony delivered on August 1, 2001, Dr. Marie Freire, then Director of the Office of Technology Transfer at NIH, stated that "[i]t is widely recognized that the Bayh-Dole Act and the Federal Technology Transfer Act continue to contribute to the global leadership of the U.S. biomedical enterprise...." An industry that was in its infancy when the Bayh-Dole Act was passed, by the end of 2005 1,415 biotechnology firms generated annual sales of $32.1 billion.[63] The number of U.S. biotechnology patents granted has increased from 619 in 1985 to 5,194 in 2006.[64]

The value of the Bayh-Dole Act might be reflected in state efforts to promote industry-university cooperation based on the contributions of these activities to local economic growth. As Mark Myers, retired Senior Vice-President of Xerox, told a meeting of the National Academy of Sciences, "[t]he role of the research university is growing ever important as an economic force in our economy...."[65] In a report for the Biotechnology Industry Organization (BIO), analysts found that there are biotechnology-related initiatives in 40 states, including many that involve cooperative efforts between academia and the private sector. Between 2000 and 2004, 19 states had developed specific bioscience strategic plans. Twenty-six states have at least one seed or venture capital program to invest in small firms undertaking work in bioscience. State laws also have been changed to allow universities to become equity partners in start up firms designed to commercialize academic R&D.[66]

CURRENT ISSUES AND CONCERNS

While the Bayh-Dole Act provides a general framework to promote expanded utilization of the results of federally funded research and development, questions have been raised as to the adequacy of current arrangements. Most experts agree that closer cooperation among government, industry, and academia can augment funding sources (both in the private and public sectors), increase technology transfer, stimulate more innovation (beyond invention), lead to new products and processes, and expand markets. However, others point out that cooperation may provide an increased opportunity for conflict of interest, redirection of research, less openness in sharing of scientific discovery, and a greater emphasis on applied rather than basic research.

The successes of the Bayh-Dole Act and the visibility of the results of its implementation have generated certain concerns, many of which are associated with the role of the university in research, as well as biomedical and biotechnology R&D, particularly as related to the availability and cost of pharmaceuticals. Several of these issues are discussed below. However, it is important to place the Bayh-Dole Act in context. The law is one significant factor in expanded industry, university, small business collaboration, but not the only one. Therefore, it may be difficult to assess what concerns are the direct result of the Bayh-Dole Act and which arise from the overall research environment. The rising costs associated with the performance of research and development, the availability of venture capital, increased R&D outsourcing by large firms, and expanded federal funding for biomedical research all contribute to increased interaction among the parties. Additional legislative initiatives including the research and experimentation tax credit, the National Cooperative Research Act, the small business technology transfer program, the advanced technology program, and cooperative R&D agreements established by the

Stevenson-Wydler Technology Innovation Act all facilitate joint R&D activities leading to the commercialization of new technologies for the marketplace.[67]

Recoupment

Over the years, several legislators have suggested that the government "recoup" its investments from firms using federally supported research and development after profits are generated. This is particularly true in the area of pharmaceuticals.[68] Such arguments are similar to those that were identified and considered as part of the original legislative debate over patent policy and cooperative R&D. The concept of recoupment is based upon the argument that the government should be reimbursed for research and development expenses provided to a contractor if the resulting product is brought to the market and generates profits. Proponents of this approach also maintained that providing the contractor with a limited time monopoly on the results of federally funded R&D through assignment of patent rights should be balanced by compensation for the government's initial investment. In the debate over related legislation, then-Senator Robert Dole stated on the floor of the Senate on April 23, 1980, the provision for recoupment was intended to insure that "the Government's investment, paid for by the taxpayers of this country, is returned to the Federal coffer."[69] During the same debate, Senator Birch Bayh argued that a payback provision means that, "in the final analysis, the taxpayer will not be out the cost of the research and they also will have the benefit of the product."[70]

Such suggestions are based on several factors. In addition to funding research performed by individual companies, under certain circumstances, the government furnishes the private sector ownership of the intellectual property resulting from this public investment. Patent protection gives firms monopoly rights on these innovations for a specified amount of time. By providing patent protection to the results of federally-funded research, a company receives an individual benefit based upon public investments. Thus, proponents of recoupment assert that the monopoly power of patents should be modified by this "public subsidization" They contend that the public has a right to a return on its investment. However, it is argued that "this right is not preserved under the patent system, which ascribes solely to the patent holder all proprietary rights and interests in the patented product or process."[71]

To date, Congress has weighed these issues and decided that, in the case of patent and technology policies, the benefits to the Nation brought about by increased innovation are paramount. The passage of the Bayh-Dole Act represented a determination that, with respect to certain types of organizations, the economic incentive to realize a return on investment provided by a patent is necessary to stimulate companies to provide the often substantial financial commitment to turn federally-funded R&D into marketable technologies and techniques. This is suggestive of the idea that the promise of a large return on investment "is precisely the tool sanctioned by the Constitution to promote the progress of science."[72] The decision was based on several determinations deriving from the rationale for federal support of basic research, the importance of technological progress to the Nation, and the critical role of private sector commercialization in technological advancement.

Federal support for basic research is founded, in large part, on the understanding that the rate of return to society as a whole generated by investments in research is significantly larger than the benefits that can be captured by any one firm performing it.[73] It has been estimated that the returns to society generated by investments in basic research are approximately twice those to the company performing the work. Government support reflects a consensus that basic research is the foundation for many innovations, but that incentives for private sector financial commitments are dampened by the fact that spending for R&D runs a high risk of failure. Even results of fruitful R&D often are exploited by other domestic and foreign companies, thus resulting in underinvestment in research by the private sector. The returns from basic research are generally long term, sometimes not marketable, and not always evident.

It is now widely accepted that "from one-third to one-half of all [U.S.] growth has come from technical progress, and that it is the principal driving force for long-term economic growth and the increased standards of living of modern industrial societies."[74] Technological advancement can clearly contribute to the resolution of those national problems which are amenable to technological solutions. Such progress is achieved through innovation, the process by which industry provides new and improved products, processes, and services. An invention becomes an innovation when it has been integrated into the economy such that the knowledge created results in a new or improved good or service that can be sold in the marketplace or is applied to production to increase productivity and quality. It is only through commercialization, a function of the business sector, that a significant stimulus to economic growth occurs. Thus, there is congressional interest in accelerating development and commercialization activities in the private sector through the Bayh-Dole Act as well as other legislation.

Actual experience and cited studies suggest that companies which do not control the results of their investments — either through ownership of patent title, exclusive license, or pricing decisions — tend to be less likely to engage in related R&D. This likelihood is reflected in the provisions of the Bayh-Dole Act (as well as other laws). Providing universities, nonprofit institutions, and small businesses with title to patents arising from federally-funded R&D offers an incentive for cooperative work and commercial application. Royalties derived from intellectual property rights provide the academic community an alternative way to support further research and the business sector a means to obtain a return on their financial contribution to the endeavor. While the idea of recoupment was considered by the Congress in hearings on the legislation, it was rejected as an unnecessary obstacle, one which would be perceived as an additional burden to working with the government. It was thought to be particularly difficult to administer.[75] Instead, Congress accepted as satisfactory the anticipated payback to the country through increased revenues from taxes on profits, new jobs created, improved productivity, and economic growth. For example, from 1980, when the Bayh-Dole Act was passed, through 2006, 5,724 new spin-off companies were created, and, in 2006 alone, 697 new products were introduced into the market by these firms.[76] The emergence of the biotechnology industry and the development of new therapeutics to improve health care are other prominent indications of such benefits. To date, these benefits have been considered more important than the initial cost of the technology to the government or any potential unfair advantage.

Government Rights: Royalty Free Licenses and Reporting Requirements

As discussed above, the government retains certain rights under the Bayh-Dole Act to protect the public interest. The act states that the government is provided a "nonexclusive, nontransferable, irrevocable, paid-up license to practice or have practiced for or on behalf of the United States any subject invention throughout the world...." This license, commonly known as a "royalty free license," has been the subject of some discussion including whether or not this permits government purchasers to obtain discounts on products developed from federally funded R&D, particularly pharmaceuticals. A July 2003 GAO report addressed this issue and concluded that the license entitles the government to practice or have practiced the invention on the government's behalf, but "does not give the federal government the far broader right to purchase, 'off the shelf' and royalty free (i.e. at a discounted price), products that happen to incorporate a federally funded invention when they are not produced under the government's license."[77] The study goes on to say that rights in one patent do not "automatically" permit rights in subsequent, related patents.[78] Because the government apparently holds few licenses on the biomedical products it purchases (generally through the Veteran's Administration and the Department of Defense),[79] federal officials indicated that procurement costs were best reduced by use of the Federal Supply Schedule and national contracts.[80] Government licenses are used primarily in the performance of research in the biomedical area.[81]

A related issue is that of tracking the government's interest in patents resulting from federally funded research and development. In an August 1999 study, GAO noted that federal contractors and grantees were not meeting the reporting requirements associated with the Bayh-Dole Act, making it difficult to identify and assess what licenses the government retained, among other things.[82] Two years later, in a follow-up report, GAO stated that four of the five agencies had taken steps to insure improved compliance with the law including several new monitoring systems, although more needed to be done.[83] Of particular interest is iEdison, created by the NIH, which electronically tracks federal inventions and is used by other agencies in addition to NIH.[84]

University Research

A question often posed is whether or not patent ownership rights provided by P.L. 96-517 have interfered with the traditional operating procedures of academia. A fear is that private sector funding of university R&D has led to conflicts of interest by scientists performing the research, particularly when academics have equity positions in the relevant companies. There are concerns that industry agendas will distort or supplant the basic research and educational responsibilities of academia. Complaints have also been expressed that the free exchange of ideas and scientific discovery are constrained as a result of both the university and the business community's interest in protecting their competitive positions.

The issue of conflict of interest is a complex one particularly when trying to determine what direct role the Bayh-Dole Act has in generating such concerns and what are the results of other factors that have lead to increased industrial funding of university research. As noted above, laws that provide tax incentives for private sector financing of university basic research and facilitate technology transfer and cooperative R&D among government, industry, and academia, as well as changes in the way companies obtain the basic research necessary for product development shape the environment within which academic research is pursued. Thus, as argued by Katherine Ku, it is necessary to evaluate criticisms of the Bayh-Dole Act and to understand that the success of the law has made many in government uncomfortable despite the clear guidelines for technology transfer it established.[85]

Senior Research Scholar Mildred Cho and her coauthors assert that the Bayh-Dole Act:

has created opportunities for conflict of interest for university faculty members because academic-industry partnerships can offer direct financial rewards to individual faculty members in the form of consulting fees, royalties, and equity in companies while simultaneously funding these faculty members' research.[86]

This, it is argued, has resulted in situations where the researcher's ties to private sector interests may not be evident and may adversely affect "the quality, outcome, and dissemination of research."[87] Other studies indicate that obligations to industry "pose a threat to scientific integrity."[88] Some commentators maintain that private sector funded research tends to generate conclusions favorable to industry; however, the factor that is primarily associated with the withholding or delay of information is the involvement of the scientist in bringing his research to market in a product, not the industrial financing itself.[89]

Data collected by Professor David Blumenthal and his colleagues also support the assessment that involvement in commercialization activities is related to delays in publication.[90] This study indicated that approximately 20% of life science researchers delayed publication of their studies more than six months at least once for reasons associated with patents and commercialization considerations. Almost 9% of faculty refused to share research or materials with other university scientists in the past three years. However, the authors conclude that "withholding of research results is not a widespread phenomenon among life-science researchers."[91] A survey of industry-university research centers by Professor Wesley Cohen and his colleagues found that over half of the centers permitted firms to request publication delays and 35% of the institutions allowed researchers to delete information prior to publication. At those centers with a mission to improve industrial products and processes, 63% allowed publication delays and 54% permitted the deletion of information.[92]

Delays in publication and the free flow of information from academia, according to Professor Richard Florida, "may well discourage or even impede the advancement of knowledge, which retards the efficient pursuit of scientific progress, in turn slowing innovation in industry."[93] Professor Florida also points to concerns over the increasing number of academic institutions taking equity positions in and/or incubating spin-off companies. These actions "simply tend to distract the university from its core missions of conducting research and generating talent." Florida concludes that publication delays and greater secrecy in the research process resulting from implementation of the Bayh-Dole Act have shifted the university away from the pursuit of its traditional goals.

Other experts, including Robert Barchi, Provost of the University of Pennsylvania, maintain that the Bayh-Dole Act has not generated a significant set of issues concerning conflicts of interest and publication delays primarily because of the importance of academic freedom to the faculty.[94] Publications are the basis for promotion and tenure and methods to respect reasonable intellectual property protection have been established. Similarly, as noted by Professor Pam Samuelson, conflicts of interest would jeopardize tenure thus regulations are in place to instruct faculty what is required of them.[95] Research conducted by Professors Pierre Azoulay, Waverly Ding, and Toby Stuart indicates that

> patenting is often accompanied by a flurry of publication activity ... academic scientists who patent are more productive than otherwise equivalent scientists that are not listed as inventors on patents, but that publication quality appears relatively similar in the two groups.[96]

In response to these issues, many universities have hired professional technology managers to work with faculty and to address patents. Universities with extensive research capabilities and resources were the first to create offices of technology transfer; after passage of the Bayh-Dole Act these offices were established with much greater frequency.[97] These university technology transfer offices have established guidelines to cover industry-university relationships, with education and publication remaining academic priorities.[98] The financial rewards derived from patenting often are only a small portion of the total amount of R&D funding for academic institutions and what substantial money does flow into individual institutions tends to be the result of one "blockbuster" patent. University technology managers report that the major reason for patent licensing is commercialization, not profit, particularly since the cost of a patent, which can run approximately $10,000, is so high.[99] While the Bayh-Dole Act focused universities on "commercially relevant technologies and closer ties between research and technological development,"[100] the costs of patenting are such that "most university licensing offices barely break even."[101]

University limitations on outside research, expeditious publication obligations mandated for certain federally-funded R&D, and conflict of interest provisions also help to preserve a balance between federal policies like the Bayh-Dole Act that promote industry-university cooperation and concerns over excessive control of the research environment by the business community. For example, NIH requires grant recipients to publish the results of their government funded R&D. This is augmented by tax code regulations necessitating prompt dissemination of actual research results in order for a university or research institution to retain its tax exempt status. NIH also has policies and guidelines promoting the availability of patents arising from federal funding for use by other scientists for research purposes without acquisition of a license.[102]

Critics argue that the Bayh-Dole Act is distorting the traditional role of the university to the detriment of future technological development. Professor Florida maintained that because universities are seen as "engines" of growth, they focus on applied rather than fundamental research. This has lead to unrealistic national and local policies and practices that encourage the commercialization of academic research while ignoring the real value of universities as the "nation's primary source of knowledge creation and talent."[103] Mildred Cho also asserted that university research is "skewed" toward marketable products and not basic research.[104] Studies by researchers Dianne Rahm and Robert P. Morgan et. al. indicated the greater the faculty interaction with industry the more the applied research.[105] According to an article in Fortune magazine, the Bayh-Dole Act has had "unintended consequences" in that "universities have evolved from public trusts into something closer to venture capital firms. What used to be a scientific community of free and open debate now often seems like a litigation scrum of data-hoarding and suspicion."[106]

Other experts disagree. A study of 3,400 faculty at six major research institutions by Professors Jerry Thursby and Marie Thursby found that "the basic/applied split in research did not change over the period 1983-1999 even though licensing had increased by a factor greater than 1 0."[107] Data collected by the National Science Foundation appear to support this assessment. According to NSF, in 1980, basic research comprised 66.6% of academic R&D endeavors while applied research and development were 33.4% of the total. In 2006, the percent of academic R&D expenditures devoted to basic research increased to 74.5% while applied research and development declined to 25.4% of the total.[108]

Commentators claim that the Bayh-Dole Act encourages the type of research that is attractive to faculty. James Severson, President of the Cornell Research Foundation, testified before the House Committee on the Judiciary that

> Today, the protection and commercialization of academic research is one way for universities to attract, retain, and reward talented faculty who wish to see the results of their research programs benefit society. A commitment to the protection of research results is important for universities to develop closer ties to companies, and to attract additional funds to support research programs.[109]

As noted by Terry Young, Assistant Vice Chancellor for Technology Transfer at Texas A&M University, the act requires funds generated by licensing to be used for future education and research necessary to "deliver 'real world' products to the public."[110] Assessing the legislation, the Biotechnology Industries Association, contends that "without the Bayh-Dole Act, few licensing agreements would be executed between private companies and federally supported research institutions, and the enormous investment our government makes in medical research would be wasted."[111]

Biotechnology and Pharmaceuticals

Many of the current concerns about the Bayh-Dole Act primarily arise out of its application to the biotechnology and pharmaceutical industries. Congressional interest in providing lower cost drugs, particularly to seniors, has focused attention on the role the act has had on the development of new pharmaceuticals for the marketplace. Certain critics maintain that the price of many therapeutics derived from federally funded R&D are excessive considering the government's financial contribution.[112] Others argue that the Bayh-Dole Act does not significantly affect pharmaceutical prices and point to a July 2001 study by NIH that found only four of the 47 FDA approved drugs generating $500 million a year were developed in part with NIH funded technologies.[113] Although the government generally does not directly support pharmaceutical research aimed at product development,[114] legislative attempts have been made to require cost controls or recoupment on drugs generated, in part, with federal funds. This is in sharp contrast to congressional and executive branch efforts, particularly in the defense arena, to make it easier for firms to acquire and utilize intellectual property associated with federally financed R&D.[115]

Overall support for biological and medical sciences has grown significantly since the passage of the Bayh-Dole Act. As measured in constant 2000 dollars, total (federal and non-federal) spending for academic R&D in these areas has increased from $4.6 billion in 1980 to $21.5 billion in 2006. Funding for university R&D in the life science, particularly biological and medical sciences, comprises by far the largest portion of academic research support. In 2006, 52% of total R&D expenditures at academic institutions went to finance the medical and biological sciences. When the Bayh-Dole Act was passed in 1980, 40.5% of the research spending at universities was in these areas.[116] While the federal government continues to be the primary source of funding for university R&D in these areas, the federal portion of academic research funding in biological sciences declined from approximately 74% in 1980 to 69% in 2006, although government support for medical research increased from approximately 64% to 66% during the same time period.[117] Expanded support for university R&D in this arena appears to be important in relation to findings by the late Professor Edwin Mansfield showing that academic research was particularly significant in the development of new products and processes in the pharmaceutical and medical device industries.[118]

Interest and activity in the biomedical and biotechnology sectors has sparked some concern over the effects of the Bayh-Dole Act on research in these areas. According to information provided by the Boston Consulting Group, in the years between 1990 and 1999, new gene patents granted increased from about 400 to 2,800 while the number granted to universities expanded from 55% to 73% during that time period.[119] Similarly, the number of U.S. biotechnology patents granted each year grew from 1,199 in 1990 to 5,194 in 2006.[120] The focus on intellectual property has led critics to charge that the Bayh-Dole Act encourages the patenting of fundamental research which, in turn, prevents further biomedical innovation. Law professors Rebecca Eisenberg and Arti Rai argue that due to the legislation, "proprietary claims have increasingly moved upstream from the end products themselves to the groundbreaking discoveries that made them possible in the first place."[121] While patents are designed to spur innovation, Rai and Eisenberg maintain that certain patents hinder the process. From their perspective, by permitting universities to patent discoveries made under federal funding, the Bayh-Dole Act "draws no distinction between inventions that lead directly to commercial products and fundamental advances that enable further scientific studies."[122] These basic innovations are generally known as "research tools."

Eisenberg and Professor Richard Nelson argue that ownership of research tools may "impose significant transaction costs" that result in delayed innovation and possible future litigation.[123] It also can stand in the way of research by others:

> Broad claims on early discoveries that are fundamental to emerging fields of knowledge are particularly worrisome in light of the great value, demonstrated time and again in history of science and technology, of having many independent minds at work trying to advance a field. Public science has flourished by permitting scientists to challenge and build upon the work of rivals.[124]

Similar concerns were expressed by Harold Varmus, President of Memorial Sloan-Kettering and former Director of NIH. In July 2000 prepared testimony, he spoke to being "troubled by widespread tendencies to seek protection of intellectual property increasingly early in the process that ultimately leads to products of obvious commercial value, because such practices can have detrimental effects on science and its delivery of health benefits."[125] While the Bayh-Dole Act and scientific advances have helped generate a dynamic biotechnology industry, there have been changes that "...are not always consistent with the best interests of science."[126]

However, as Varmus and others acknowledge, the remedies to this situation are not necessarily associated with the Bayh-Dole Act. Yale's Richard Levin noted that while some research should be kept in the public domain, including research tools, the fact that it is privatized is not the result of the Bayh-Dole Act, but rather the result of patent law made by the courts and the Congress. Therefore, he believes that changes to the act are not the appropriate means to address the issues.[127] Other experts agree that "many of the issues that are identified today as negative consequences of Bayh-Dole can be traced to the institutional polices [of universities] structured to optimize institutional benefits and income, rather than to the Act itself."[128]

Current law, as reaffirmed by court decisions, permits the patenting of research tools. However, there have been efforts to encourage the widespread availability of these tools. Marie Freire testified that the value to society is greatest if the research tools are easily available for use in research. She asserted that there is a need to balance commercial interests with public interests.[129] To achieve this balance, NIH has developed guidelines for universities and companies receiving federal funding that make clear research tools are to be made available to other scientists under reasonable terms.[130] In addition, the U.S. Patent and Trademark Office recently made changes in the guidelines used to determine the patentability of biotechnology discoveries.

Studies by Professors John Walsh, Ashish Arora, Wesley Cohen, and Charlene Cho found that although there are now more patents associated with biomedical research, and on more fundamental work, there is little evidence that work has been curtailed due to intellectual property issues associated with research tools.[131] Scientists are able to continue their research by "licensing, inventing around patents, going offshore, the development and use of public databases and research tools, court challenges, and simply using the technology without a license (i.e., infringement)." According to the authors, private sector owners of patents permitted such infringement in academia (with the exception of those associated with diagnostic tests in clinical trials) "partly because it can increase the value of the patented technology."

CONCLUDING OBSERVATIONS

The discussion surrounding changes to the patent laws in the 1980s, and the debate over technology transfer since the late 1970s, acknowledged many of the issues currently being explored. As a result of expressed concerns, certain safeguards were built into the activities authorized by the Bayh-Dole Act. As discussed previously, march-in rights, the government's retention of an irrevocable license to patents generated under federally funded R&D, publication requirements, and commercialization schedules, among other things, all are incorporated into the process to protect the public interest. While there is a potential for creating an "unfair" advantage for one company over another, this is balanced against the need for new technologies and techniques and their contribution to the well-being of the Nation.

Despite arguments that title should remain in the public sector where it is accessible to all interested parties, the earlier lack of exclusivity appeared to interfere with the further development and commercialization of federally funded R&D. During the 1980s, Congress determined that the dispensation of patent rights to universities, small businesses, and nonprofit institutions and cooperative efforts took precedence, projecting the greater good generated by new products and processes that improve the country's health and welfare. Lawmakers anticipated the economic benefits through increased revenues from profits, wages, and salaries. The government receives a significant payback through taxes on profits and society benefits from new jobs created and expanded productivity. The importance of patent ownership has been reinforced by the positive effects studies have demonstrated P.L. 96-517 is reported to have had on the emergence of new technologies and new techniques generated by American companies.

There remain areas of concern, as discussed above, that Congress may decide to pursue. Some argue, particularly with respect to pharmaceuticals and biotechnology, that under the Bayh-Dole Act companies are receiving too many benefits at the expense of the public. Others, particularly in the defense arena, assert that the existing rights retained by the government under the act are too restrictive and are an impediment to meeting federal needs. But the impact of the legislation is still seen as significant. As summed up by Howard Bremer, who was patent counsel to the Wisconsin Alumni Research Foundation from 1960 through 1988:

> One important factor, which is often overlooked, is that the success was achieved without cost to the taxpayer. In other words, no separate appropriation of government funds was needed to establish or manage the effort. In fact, it has been estimated that the economic benefits flowing from the

universities' licensing activities adds about $41 billion to the United States economy.

Significant as that dollar amount is, it should not be overlooked that university inventions, arising, as most of them do, from basic research, have led to many products which have or exhibit the capability of saving lives or of improving the lives, safety and health of the citizens of the United States and around the world. In that context their contribution to society is immeasurable.[132]

REFERENCES

[1] For additional discussion see CRS Report RL33528, Industrial Competitiveness and Technological Advancement: Debate Over Government Policy, by Wendy H. Schacht.

[2] National Science Board, Science and Engineering Indicators — 2006, Washington, National Science Foundation, A4-5.

[3] House Committee on the Judiciary, Report to Accompany H.R. 6933, 96th Cong., 2d Sess., H.Rept. 96-1307, Part 1, 3.

[4] For more information see CRS Report RL32324, Federal R&D, Drug Discovery, and Pricing: Insights from the NIH-University-Industry Relationship, by Wendy H. Schacht.

[5] See Frederic M. Sherer, Industrial Market Structure and Economic Performance (1970), 384-87.

[6] See John R. Thomas, "Collusion and Collective Action in the Patent System: A Proposal for Patent Bounties," University of Illinois Law Review (2001), 305.

[7] Ibid.

[8] Federic M. Scherer, "The Economics of Human Gene Patents," 77 Academic Medicine, December 2002, 1350.

[9] Wesley M. Cohen, Richard R. Nelson, and John P. Walsh, Protecting Their Intellectual Assets: Appropriability Conditions and Why U.S. Manufacturing Firms Patent (or Not), NBER, February 2000, available at [http://www.nber.org/papers/w7552].

[10] Richard C. Levin, Alvin K. Klevorick, Richard R. Nelson, and Sidney G. Winter, "Appropriating the Returns for Industrial Research and Development," Brookings Papers on Economic Activity, 1987, printed in The Economics of Technical Change, Edwin Mansfield and Elizabeth Mansfield, eds., (Vermont, Edward Elgar Publishing Co., 1993), 253.

[11] Protecting Their Intellectual Assets: Appropriability Conditions and Why U.S. Manufacturing Firms Patent (or Not), Table 1.
[12] "Innovation's Golden Goose," The Economist (US), December 14, 2002.
[13] National Science Board, Science and Engineering Indicators — 2002, Washington, National Science Foundation, A4-9.
[14] National Science Board, Science and Engineering Indicators — 1993, Washington, National Science Foundation, 185.
[15] National Science Board, Science Indicators — 1976, Washington, National Science Foundation, 116.
[16] Eleanor H. Erdevig, "Small Business, Big Job Growth," Chicago Economic Perspectives, November-December 1986, 22.
[17] Marc Levinson, "Small Business: Myth and Reality," Dun's Business Month, September 1985, 32-33.
[18] Zoltan J. Acs and David B. Audretsch, Innovation and Small Firms (Cambridge: The MIT Press, 1990), 50-5 1.
[19] Charles Brown, James Hamilton, and James Medoff, Employers Large and Small, (Cambridge: Harvard University Press, 1990), 10.
[20] William J. Baumol, Education for Innovation: Entrepreneurial Breakthroughs vs. Corporate Incremental Improvements, NBER, June 2004, 2-3, available at [http://www.nber.org/papers/10578].
[21] National Academy of Engineering, Risk & Innovation, The Role and Importance of Small High-Tech Companies in the U.S. Economy (Washington: National Academy Press, 1995), 37.
[22] David R. King, Jeffrey G. Covin, W. Harvey Hegarty, "Complementary Resources and the Exploitation of Technological Innovations," Journal of Management, August 1, 2003, 592.
[23] P.L. 96-5 17, sec. 200.
[24] Ibid.
[25] Report to Accompany H.R. 6933, 5.
[26] P.L. 96-517, sec. 203.
[27] Senate Committee on the Judiciary, Report to Accompany S. 2171, 98th Cong., 2d Sess. S.Rept. 98-662, 1984, 3.
[28] For additional discussion see Industrial Competitiveness and Technological Advancement: Debate Over Government Policy.
[29] P.L. 96-5 17, sec. 200.
[30] U.S. General Accounting Office, Patent Policy: Recent Changes in Federal Law Considered Beneficial, RCED-87-44, April 1987, 3.
[31] Ibid., 3.

[32] National Science Foundation, "Changes in Federal and Non-Federal Support for Academic R&D Over the Past Three Decades," InfoBrief, June 2002 available at [http://www.nsf.gov], National Science Foundation, "National Patterns of R&D Resources: 2003, Special Report," available at [http://www.nsf.gov/statistics/nsf05308/pdfstart.htm], and National Science Foundation, "Universities Report Stalled Growth in Federal R&D Funding in FY2006," InfoBrief, September 2007, available at [http://www.nsf.gov/statistics/ infbrief/nsf07336/nsf07336.pdf].
[33] Patent Policy: Recent Changes in Federal Law Considered Beneficial, 20-21.
[34] Ibid., 16.
[35] Ibid., 23.
[36] U.S. General Accounting Office, Technology Transfer: Administration of the Bayh-Dole Act by Research Universities, RCED-98-126, May 1998, 2.
[37] Ibid., 15.
[38] National Academy of Sciences, Board on Science, Technology, and Economic Policy, Workshop on Academic IP: Effects of University Patenting and Licensing on Commercialization and Research, April 17, 2001 [transcript], 261-262 available at [http://www.nas.edu].
[39] Innovation's Golden Goose
[40] Workshop on Academic IP: Effects of University Patenting and Licensing on Commercialization and Research, 9.
[41] Science and Engineering Indicators — 1993, 430.
[42] U.S. Patent and Trademark Office, Utility Patents Assigned to U.S. Colleges and Universities, available at [http://www.uspto.gov/web/offices/ac/ido/oeip/taf/univ/asgn/table_ 1_2005.htm].
[43] Association of University Technology Managers, U.S. AUTM Licensing Survey: FY2006, available at [http://www.autm.net/events/file/ AUTM_06_US%20LSS_FNL.pdf].
[44] Goldie Blumenstyk, "Income From University Licenses on Patents Exceeded $1-Billion," The Chronicle of Higher Education, March 22, 2002.
[45] Harun Bulut and GianCarlo Moschini, U.S. Universities' Net Returns from Patenting and Licensing: A Quantile Regression Analysis, Iowa State University Working Paper-06-WP 432, September 2006, 2 available at [http://www.card.iastate.edu].

[46] Gregory K. Sobolski, John H. Barton, Ezekiel J. Emanuel, "Technology Licensing, Lessons From the US Experience," Journal of the American Medical Association, December 28, 2005, 3138.
[47] Association of University Technology Managers, AUTM Licensing Survey: FY2004, available at [http://www.autm.net/about/dsp.pub Detail2.cfm?pid=28].
[48] David C. Mowery, Richard R. Nelson, Bhaven N. Sampat, and Arvids A. Ziedonis, "The Growth of Patenting and Licensing by U.S. Universities: An Assessment of the Effects of the Bayh-Dole Act of 1980," Research Policy 30, 2001, 99.
[49] Ibid., 100.
[50] David C. Mowery, The Bayh-Dole Act and High-Technology Entrepreneurship in U.S. Universities: Chicken, Egg, or Something Else?, paper prepared for the Eller Center conference on "Entrepreneurship Education and Technology Transfer," University of Arizona, January 21-22, 2005, available at [http://entrepreneurship.eller.arizona.edu/docs/conferences/2005/colloquium/D_Mowery.pdf].
[51] Workshop on Academic IP: Effects of University Patenting and Licensing on Commercialization and Research, 17.
[52] Science and Engineering Indicators — 1993, 152.
[53] Ibid., 152 and Workshop on Academic IP: Effects of University Patenting and Licensing on Commercialization and Research, 57-58.
[54] Ashley J. Stevens, "Is Bayh-Dole Under Siege Again?" Technology Access Report, July 2001. See also Lori Turk-Bicakci and Steven Brint, "University-Industry Collaboration: Patterns of Growth for Low- and Middle-Level Performers," Higher Education, January 2005, 61-89.
[55] Bhavan N. Sampat, "Patenting and US Academic Research in the 20th Century: The World Before and After Bayh-Dole," Research Policy, July 2006, 773.
[56] Ibid, 773.
[57] Marie Thursby, Jerry Thursby, and Emmanuel Dechenaux, Shirking, Shelving, and Sharing Risk: The Role of University License Contracts, April 9, 2004, National Bureau of Economic Research, available at [http://www.nber.org/~confer/2004/entf04/thursby.pdf].
[58] Jerry G. Thursby and Marie C. Thursby, University Licensing Under Bayh-Dole: What are the Issues and Evidence?, May 2003, available at [http://opensource.mit.edu/papers/Thursby.pdf].
[59] Ibid.

[60] Scott Shane, "Encouraging University Entrepreneurship? The Effect of the Bayh-Dole Act on University Patenting in the United States," Journal of Business Venturing 19, 2004, 128.
[61] Patent Policy: Recent Changes in Federal Law Considered Beneficial, 4.
[62] House Committee on the Judiciary, Biotechnology Development and Patent Law, 102d Cong., 1st Sess., November 20, 1991, 48.
[63] Biotechnology Industry Organization, Biotechnology Industry Facts, available at [http://www.bio.org/speeches/pubs/er/statistics.asp].
[64] National Science Foundation, Science and Engineering Indicators, 2008, Appendix tables 6-48 and 6-49, available at [http://www.nsf.gov/statistics/seind08/append/c6/at06-48pdf] and [http://www.nsf.gov/statistics/seind08/append/c6/at06-49pdf].
[65] Workshop on Academic IP: Effects of University Patenting and Licensing on Commercialization and Research, 255.
[66] Battelle Technology Partnership Practice and SSTI, Laboratories of Innovation: State Bioscience Initiatives 2004, June 2004, 27-29, available at [http://www.mdbio.org/pdf/ reports/2004_bioscience_initiatives. pdf?MDBIOSESSION=].
[67] For additional information see CRS Report RL3 3526, Cooperative R&D: Federal Efforts to Promote Industrial Competitiveness, by Wendy H. Schacht, and CRS Report RL33528, Industrial Competitiveness and Technology Advancement: Debate Over Government Policy, by Wendy H. Schacht.
[68] For a more detailed discussion of this issue in the pharmaceutical arena see CRS Report RL32324, Federal R&D, Drug Discover, and Pricing: Insights from the NIH-UniversityIndustry Relationship, by Wendy H. Schacht.
[69] U.S. Congress, Congressional Record, April 23, 1980, S739.
[70] Ibid, S743.
[71] Steven R. Salbu, "Aids and Drug Pricing: In Search of a Policy," Washington University Law Quarterly, Fall 1993, 5-20.
[72] Evan Ackiron, "Patents for Critical Pharmaceuticals: The AZT Case," American Journal of Law and Medicine, 1991, 18.
[73] Edwin Mansfield, "Social Returns From R&D: Findings, Methods, and Limitations," Research/Technology Management, November-December 1991, 24.
[74] Gregory Tassey, The Economics of R&D Policy (Connecticut: Quorum Books, 1997), 54. See also: Edwin Mansfield, "Intellectual Property Rights, Technological Change, and Economic Growth," in: Intellectual

Property Rights and Capital Formation in the Next Decade, eds. Charls E. Walker and Mark A. Bloomfield (New York: University Press of America, 1988), 5.
[75] For example see U.S. House of Representatives, Committee on Science and Technology, Government Patent Policy, Hearings, September 23, 27, 28, 29, and October 1, 1976, 94th Cong. 2nd sess., 1976; United States Senate, Select Committee on Small Business, Government Patent Policies, Hearings, December 19, 20, and 21, 1977, 95th Cong. 1st sess., 1978; and U.S. Senate, Committee on Commerce, Science, and Transportation, Patent Policy, Hearings, July 23 and 27, and October 25, 1979, 96th Cong. 1st sess., 1979.
[76] U.S. AUTM. Licensing Survey: FY2006, 5.
[77] General Accounting Office, Technology Transfer: Agencies' Rights to Federally Sponsored Biomedical Inventions, GAO-03-536, July 2003, 7.
[78] Ibid., 8.
[79] Ibid., 8.
[80] Ibid., 12.
[81] Ibid., 10.
[82] General Accounting Office, Technology Transfer: Reporting Requirements for Federally Sponsored Inventions Need Revision, August 1999, GAO/RCED-99-242, 2.
[83] General Accounting Office, Intellectual Property: Federal Agency Efforts in Transferring and Reporting New Technology, October 2002, GAO-03-47, 29.
[84] Ibid., 33.
[85] Workshop on Academic IP: Effects of University Patenting and Licensing on Commercialization and Research, 98, 100-101.
[86] Mildred K. Cho, Ryo Shohara, Anna Schissel, and Drummond Rennie, "Policies on Faculty Conflicts of Interest at U.S. Universities," Journal of the American Medical Association, November 1, 2000.
[87] Ibid.
[88] Justin E. Bekelman, Yan Li, and Cary P. Gross, "Scope and Impact of Financial Conflicts of Interest in Biomedical Research: A Systematic Review," Journal of the American Medical Association, January 22/January 29, 2003.
[89] Ibid.
[90] David Blumenthal, Eric G. Campbell, Melissa S. Anderson, Nancyanne Causino, and Karen Seashore Louis, "Withholding Research Results in Academic Life," Journal of the American Medical Association, April 16, 1997, 1224.

[91] Ibid., 1224
[92] Wesley M. Cohen, Richard Florida, Lucien Randazzese, and John Walsh, "Industry and the Academy: Uneasy Partners in the Cause of Technological Advance," in: Challenges to Research Universities, eds. Linda R. Cohen, Wesley Cohen, Roger Noll, William Rogerson, and Albert Teich (Washington: The Brookings Press, 1998), 188-189.
[93] Richard Florida, "The Role of the University: Leveraging Talent, Not Technology," Issues in Science and Technology, Summer 1999.
[94] Workshop on Academic IP: Effects of University Patenting and Licensing on Commercialization and Research, 19-20.
[95] Workshop on Academic IP: Effects of University Patenting and Licensing on Commercialization and Research, 193.
[96] Pierre Axoulay, Waverly Ding, and Toby Stuart, The Impact of Academic Patenting on the Rate, Quality, and Direction of (Public) Research, National Bureau of Economic Research, January 2006, available at [http://www.nber.org/papers/w11917.pdf].
[97] Everett Rogers, Jing Yin, and Joern Hoffmann, "Assessing the Effectiveness of Technology Transfer Offices at U.S. Research Universities," Journal of the Association of University Technology Managers, v. XII, 2000, available at [http://www.autm.net].
[98] Technology Transfer: Administration of the Bayh-Dole Act by Research Universities.
[99] Ann M. Thayer, "University Technology Moves to Market via Patenting, Licensing,: Chemical and Engineering News, August 24, 1992, 17-18. See also: Jerry G. Thursby and Marie C. Thursby, "Intellectual Property: University Licensing and the Bayh-Dole Act," Science, August 22, 2003, 1052.
[100] National Science Board, Science and Engineering Indicators - - 2002, Washington, 5-54.
[101] Lita Nelson, "Increase of Intellectual Property Licensing at Universities Stems from Changes in Funding and Legislation," MIT Tech Talk, August 26, 1998, available at [http://web.mit.edu].
[102] Available at [http://www.nih.gov].
[103] The Role of the University: Leveraging Talent, Not Technology.
[104] Eric Niller, "Biotech & Health: Report Fails to Address the Downside of Academic- Industry Collaborations," Wall Street Journal (Europe), August 6, 2001, 17.
[105] Industry and the Academy: Uneasy Partners in the Cause of Technological Advance, 186.

[106] Clifton Leaf, "The Law of Unintended Consequences," Fortune, September 19, 2005, 252.
[107] University Licensing Under Bayh-Dole: What are the Issues and Evidence?
[108] Science and Engineering Indicators, 2008, Appendix table 5-1, available at [http://www.nsf.gov/statistics/seind08/append/c5/at05-01.pdf].
[109] House Committee on the Judiciary, Subcommittee on Courts and Intellectual Property, Hearings on Gene Patents and Other Genomic Inventions, July 13, 2000, available at [http://www.house.gov/judiciary/seve07 13 .htm].
[110] U.S. Department of Commerce, Technology Administration, Innovation in America: University R&D, June 11, 2002, available at [http://www.ta.doc.gov/reports].
[111] Biotechnology Industry Organization, Testimony on Bayh-Dole and Technology Transfer Before the President's Council on Science and Technology, Office of Science and Technology Policy, April 11, 2002, available at [http://www.bio.org].
[112] See CRS Report RL32324, Federal R&D, Drug Discovery, and Pricing: Insights from the NIH-University-Industry Relationship, by Wendy H. Schacht.
[113] Department of Health and Human Services, National Institutes of Health, NIH Response to the Conference Report Request for a Plan to Ensure Taxpayers' Interests are Protected, July 2001, available on the web at [http://www.nih.gov/news/070 10 1wyden.htm].
[114] See CRS Report RL309 13, Pharmaceutical Research and Development: A Description and Analysis of the Process, by Richard E. Rowberg.
[115] See House Committee on Government Reform, Subcommittee on Technology and Procurement Policy, Toward Greater Public-Private Collaboration in Research and Development: How the Treatment of Intellectual Property Rights is Minimizing Innovation in the Federal Government, hearings, July 17, 2001, available at [http://www.house.gov/reform].
[116] Science and Engineering Indicators, 2008, Appendix table 5-4, available at [http://www.nsf.gov/statistics/seind08/append/c5/at05-04.pdf].
[117] Science and Engineering Indicators, 2008, Appendix table 5-3, available at [http://www.nsf.gov/statistics/seind08/append/c5/at05-03.pdf].

[118] Edwin Mansfield, "Academic Research and Industrial Innovation: An Update of Empirical Findings," Research Policy, 1998, 773-776.
[119] Hamilton Moses, III and Joseph B. Martin, "Academic Relationships with Industry," Journal of the American Medical Association, February 21, 2001, 933.
[120] Science and Engineering Indicators, 2008, Appendix Tables 6-48 and 6-49, available at [http://www.nsf.gov/statistics/seind08/append/c6/at06-48.pdf] and [http://www.nsf.gov/ statistics/seind08/append/c6/at06-49.pdf].
[121] Arti K. Rai and Rebecca S. Eisenberg, "Bayh-Dole Reform and the Progress of Biomedicine," American Scientist, January- February 2003, 52.
[122] Ibid.
[123] Rebecca S. Eisenberg and Richard R. Nelson, "Public vs. Proprietary Science: A Fruitful Tension?," Daedalus, Spring 2002.
[124] Ibid.
[125] Hearings on Gene Patents and Other Genomic Inventions.
[126] Ibid.
[127] Workshop on Academic IP: Effects of University Patenting and Licensing on Commercialization and Research, 262.
[128] Sara Boettiger and Alan B. Bennet, "Bayh-Dole: IF We Knew Then What We Know Now," Nature Biotechnology, 24 (2006), available at [http://www.nature.com/nbt/journal/v24/n3/full/nbt0306-320.html].
[129] Senate Committee on Appropriations, Subcommittee on Labor, Health and Human Services, Education and Related Agencies, Hearings, August 1, 2001.
[130] Available on the NIH website at [http://www.nih.gov].
[131] John P. Walsh, Ashish Arora, Wesley M. Cohen, "Working Through the Patent Problem," Science, February 14, 2003, 1021 and John P. Walsh, Charlene Cho, and Wesley Cohen, "View for the Bench: Patents and Material Transfers," Science, September 23, 2005, 2002-2003.
[132] Howard Bremer, "The First Two Decades of the Bayh-Dole Act as Public Policy," National Association of State Universities and Land-Grant Colleges, November 11, 2001, available at [http://www.nasulgc.org].

INDEX

A

academic, viii, ix, 14, 15, 19, 30, 34, 37, 39, 40, 56, 59, 60, 64, 65, 66, 68, 71, 72, 73, 74, 75, 76
academics, 72
access, 26, 32, 34, 42, 47, 53
ACI, 3
ad hoc, vii, 1, 5
administrators, 64
aerospace, 46, 59
agricultural, 22, 47
alternative, 10, 40, 71
amendments, viii, 17, 30, 51
American Competitiveness Initiative, 3, 8, 31
analysts, 57, 65, 66, 68
antitrust, 2, 8, 15, 16, 22, 32, 35, 36, 47
antitrust laws, 2, 8, 15, 35, 47
application, 4, 11, 16, 30, 31, 36, 39, 43, 44, 49, 57, 58, 63, 71, 76
applied research, 5, 23, 33, 49, 52, 75
appropriations, 2, 3, 10, 11, 12, 13, 19, 20, 21, 24, 25, 26, 30, 37, 38, 39, 48, 53
argument, 69
Arizona, 83
Asian, 44
assessment, 66, 73, 75
assignment, 69
assumptions, 7
ATP, 2, 11, 12, 13, 27, 30, 36, 37, 38, 39, 48

authority, 17
availability, 19, 35, 44, 46, 62, 68, 75, 78
awareness, 17

B

barrier, 14, 59
barriers, 9, 26, 35, 53, 57
basic research, ix, 2, 3, 5, 8, 9, 14, 15, 31, 32, 34, 39, 47, 56, 58, 60, 68, 70, 72, 75, 80
beliefs, 6
beneficial effect, 67
benefits, ix, 4, 5, 7, 33, 43, 56, 57, 59, 64, 69, 70, 71, 78, 79
biotechnology, ix, 5, 33, 56, 57, 66, 67, 68, 71, 76, 77, 78, 79
bipartisan, 17, 22
bonds, 3, 24, 26, 31, 52, 53
Boston, 77
bureaucracy, 58
Bush Administration, 8, 12, 20, 38

C

capacity, 34, 61
capital intensive, 61
catalyst, 66
channels, 66
children, 8
citizens, 80

civilian, 8, 46
clinical trial, 78
clinical trials, 78
Clinton Administration, 7, 11, 20, 37, 42
Co, 41, 60, 80
collaboration, ix, 33, 34, 35, 42, 56, 58, 60, 62, 64, 66, 68, 83, 87
Columbia, 65, 66
Columbia University, 65, 66
commercialization, ix, 4, 5, 7, 8, 9, 10, 14, 16, 17, 18, 19, 33, 34, 35, 40, 41, 42, 44, 46, 55, 56, 57, 59, 60, 62, 63, 65, 66, 67, 69, 70, 73, 74, 75, 79
Committee on Appropriations, 2, 12, 13, 21, 25, 27, 38, 39, 48, 50, 88
Committee on the Judiciary, 26, 53, 75, 80, 81, 84, 87
communities, 14, 34, 60
community, ix, 7, 10, 14, 15, 16, 19, 31, 33, 35, 40, 41, 42, 45, 55, 57, 58, 60, 71, 72, 74, 75
comparative advantage, 4, 31
compensation, 59, 63, 69
competition, viii, 4, 11, 22, 30, 31, 32, 35, 36, 45, 59
competitiveness, vii, viii, ix, 1, 2, 3, 6, 7, 14, 29, 30, 31, 32, 33, 41, 43, 44, 45, 47
compliance, 72
components, 6, 7
computer software, 35
concentration, 33, 59
conception, 33
conflict, ix, 56, 68, 72, 73, 74
conflict of interest, ix, 56, 68, 72, 73, 74
Congress, iv, vii, viii, ix, 1, 2, 3, 6, 10, 11, 12, 13, 15, 17, 20, 21, 22, 23, 27, 29, 30, 31, 37, 38, 39, 43, 47, 48, 51, 57, 58, 62, 64, 69, 71, 78, 79, 84
Congressional Record, 84
Connecticut, 84
consensus, vii, 1, 5, 70
consent, 26, 48, 53
Consolidated Appropriations Act, 10, 11, 20, 21, 37, 38
Constitution, 58, 70

constraints, 19
construction, 24, 25
consulting, 73
consumers, 44
contractors, 15, 16, 18, 40, 63, 72
contracts, 71
control, ix, 56, 70, 74
corporations, 18, 61
costs, viii, ix, 9, 15, 29, 31, 32, 44, 45, 55, 59, 60, 65, 68, 71, 74, 77
costs of production, 45
counsel, 79
courts, 78
credit, 3, 9, 14, 15, 31, 39
CRS, 1, 11, 14, 15, 16, 17, 22, 27, 29, 37, 40, 41, 42, 43, 46, 48, 55, 80, 84, 87

D

database, 24, 25
decisions, ix, 6, 7, 18, 42, 43, 47, 56, 64, 70, 78
deduction, 14, 39
defense, 35, 40, 46, 57, 58, 76, 79
definition, 47
delivery, 44, 78
demand, 46
Department of Commerce, viii, 2, 6, 8, 11, 36, 40, 87
Department of Defense, 17, 19, 40, 41, 45, 71
Department of Energy, 8, 40, 45, 46, 63
Department of Health and Human Services, 87
Department of Justice, 16
desire, 59
development policy, 9
differentiation, 5, 33
diffusion, 5, 44
direct measure, 2, 22
disclosure, 16, 32, 36, 59, 63
discounts, 71
Discover, 84
Discovery, 80, 87
discretionary, 18
dislocations, 7

disseminate, 17
distribution, 4, 31
divergence, 7
diversity, 62
DNA, 36
donations, 14, 39
drugs, 76

E

East Asia, 44
economic development, 42, 50, 51, 65
economic growth, vii, 1, 4, 6, 7, 8, 9, 31, 35, 67, 70, 71
economic policy, 7
Economic Recovery Tax Act, 9, 14, 39
economic security, 40
Education, 3, 24, 26, 31, 49, 50, 52, 53, 81, 82, 83, 88
educational process, 14, 34
electric power, 35
employees, 17, 61
employment, 61
energy, 3, 8, 23, 31, 51
engines, 75
enterprise, 5, 35, 41, 58, 67
entrepreneurship, 83
environment, 6, 7, 22, 33, 34, 57, 68, 72, 74
equity, 68, 72, 73
Europe, 86
Executive Order, 8
expenditures, 9, 14, 39, 75, 76
experimentation tax credit, 3, 26, 27, 31, 39, 48, 49, 50, 53, 54, 68
expert, iv
expertise, viii, 7, 15, 16, 17, 29, 30, 32, 33, 34, 45, 47, 60
exploitation, 59

F

failure, 70
faith, 16
FDA, 76

fear, 6, 72
February, 7, 13, 21, 26, 39, 50, 53, 63, 80, 88
federal budget, 42
federal funds, 5, 42, 43, 76
federal government, vii, viii, 2, 4, 8, 13, 14, 17, 19, 20, 31, 32, 34, 41, 43, 45, 57, 60, 64, 65, 71, 76
Federal Register, 19
Federal Reserve, 61
Federal Reserve Bank, 61
fee, 16, 36
fees, 16, 73
finance, 2, 12, 21, 35, 58, 76
financial support, 12, 32, 34, 45, 60
financing, 11, 23, 36, 37, 38, 46, 47, 48, 49, 50, 51, 52, 64, 72, 73
firm size, 61
firms, 3, 4, 10, 11, 13, 18, 22, 30, 32, 33, 35, 37, 40, 43, 45, 47, 58, 59, 61, 62, 63, 65, 67, 68, 69, 71, 73, 75, 76
flat panel displays, 22, 47
flexibility, 32
flow, 47, 73, 74
foreign firms, 5, 31, 47
foreign person, 16, 36
freedom, 74
funding, vii, viii, ix, 1, 2, 3, 8, 9, 10, 11, 12, 13, 14, 19, 21, 22, 23, 24, 25, 30, 32, 33, 34, 35, 36, 37, 38, 39, 43, 45, 47, 49, 52, 56, 60, 63, 64, 66, 68, 69, 72, 73, 74, 75, 76, 77, 78
funds, 11, 15, 18, 32, 34, 35, 37, 40, 43, 44, 67, 75, 76, 79

G

GAO, 64, 67, 71, 72, 85
gene, 77
General Accounting Office, 64, 81, 82, 85
generation, viii, 4, 29, 41, 61
global leaders, 67
goals, 8, 22, 34, 64, 73
goods and services, ix, 5, 34, 44, 45, 56, 60, 62

government, vii, viii, ix, 2, 3, 4, 5, 6, 7, 8, 9, 11, 13, 14, 15, 17, 18, 19, 20, 22, 23, 29, 30, 31, 32, 34, 35, 36, 40, 41, 42, 43, 44, 45, 47, 51, 55, 56, 57, 58, 60, 62, 63, 64, 65, 66, 68, 69, 71, 72, 74, 76, 79
Government Accountability Office, 64
government intervention, 6
government policy, 7, 43
grants, vii, ix, 3, 9, 10, 12, 13, 19, 23, 24, 25, 31, 37, 42, 49, 50, 51, 52, 55, 67
gross national product, 61
grounding, 8
groups, 7, 74
growth, vii, 1, 4, 6, 7, 8, 9, 31, 35, 61, 66, 67, 70, 71, 75
guidelines, 32, 43, 72, 74, 75, 78

H

Harvard, 81
health, 8, 35, 61, 63, 71, 78, 79, 80
Health and Human Services, 87, 88
health care, 71
health insurance, 61
healthcare, 36
high risk, 70
high tech, 8, 43, 61
higher education, 14, 39
high-risk, 13
high-tech, 7, 9, 61
hip, 67
hips, 44
hiring, 32
House, 2, 3, 10, 11, 12, 13, 15, 20, 21, 23, 24, 25, 26, 31, 37, 38, 39, 48, 49, 50, 51, 52, 53, 57, 62, 75, 80, 84, 85, 87
House Committee on Government Reform, 87
human, 65
human resources, 65
humanity, 65

I

id, 12, 38, 64

Illinois, 80
implementation, 17, 57, 65, 68, 73
in situ, 73
incentive, ix, 18, 55, 56, 59, 65, 67, 70
incentives, 2, 7, 9, 17, 22, 35, 41, 44, 47, 58, 59, 66, 70
income, 5, 33, 65, 78
indication, 42
indirect measure, 2, 22
industrial, vii, ix, 4, 6, 7, 8, 14, 15, 30, 31, 33, 34, 36, 41, 45, 59, 60, 70, 72, 73
industrial policy, 6, 7
industrial sectors, 4, 30, 31, 34, 36, 45, 59
industrialized countries, 7
industry, vii, viii, ix, 2, 4, 5, 6, 7, 8, 9, 13, 14, 15, 16, 17, 18, 19, 21, 22, 29, 30, 31, 32, 33, 34, 35, 39, 40, 41, 42, 43, 44, 45, 46, 47, 55, 56, 57, 58, 59, 60, 61, 62, 64, 66, 67, 68, 70, 71, 72, 73, 74, 75, 78
infancy, 67
information technology, 22
infrastructure, 8, 19, 36
infringement, 78
injury, iv
innovation, viii, ix, 3, 4, 5, 6, 7, 8, 9, 10, 13, 14, 17, 18, 22, 23, 24, 25, 26, 29, 30, 31, 33, 34, 35, 41, 42, 44, 49, 50, 51, 53, 55, 56, 57, 58, 59, 61, 62, 64, 65, 66, 67, 68, 69, 70, 73, 77, 81, 82, 84, 87, 88
institutions, 3, 9, 14, 15, 18, 21, 23, 31, 34, 37, 39, 41, 49, 51, 52, 58, 60, 63, 64, 70, 73, 74, 75, 76, 79
insurance, 61
integration, 6
integrity, 73
Intel, 46
intellectual property, ix, 2, 8, 17, 41, 47, 56, 57, 60, 65, 66, 69, 71, 74, 76, 77, 78
intellectual property rights, 17, 47, 71
interaction, 15, 18, 40, 47, 64, 68, 75
interactions, 66
interference, 6, 7
Internal Revenue Code, 26, 48, 53
international trade, 4, 31
internationalization, 7

internet, 19
interrelationships, 6
intervention, 6
interviews, 64
inventions, vii, viii, ix, 4, 15, 17, 18, 30, 32, 40, 41, 49, 55, 56, 58, 60, 62, 63, 64, 66, 67, 72, 77, 80
inventors, 18, 74
investment, ix, 6, 8, 11, 17, 22, 30, 36, 55, 58, 59, 60, 69, 70, 76
IP, 82, 83, 84, 85, 86, 88
isolation, 14

J

January, 23, 26, 37, 49, 50, 51, 52, 53, 83, 85, 86, 88
Japan, 44
job creation, 8
jobs, 4, 31, 61, 71, 79
joint ventures, vii, viii, ix, 13, 15, 16, 29, 30, 32, 36, 40, 41, 43, 44, 45, 46, 57
Judiciary, 26, 49, 50, 53, 87
jurisdiction, 5
jurisdictions, 42
Justice Department, 36

K

King, 81

L

labor, 4, 62
language, 25
large-scale, 45
law, 9, 10, 11, 15, 16, 17, 18, 20, 32, 36, 40, 42, 48, 56, 57, 58, 62, 63, 64, 68, 72, 78
laws, viii, 2, 8, 14, 15, 22, 30, 35, 39, 40, 41, 47, 57, 60, 68, 70, 72, 79
lead, ix, 33, 45, 46, 56, 59, 68, 72, 75, 77
leadership, 22, 44
learning, 59

legislation, ix, 9, 11, 12, 13, 15, 19, 20, 21, 30, 35, 37, 38, 39, 48, 56, 57, 60, 61, 62, 64, 65, 66, 69, 70, 71, 76, 77, 79
legislative, viii, 2, 18, 33, 35, 40, 41, 47, 57, 63, 68, 69, 76
licenses, 18, 19, 63, 65, 71, 72
licensing, 15, 32, 40, 59, 63, 64, 65, 66, 67, 74, 75, 76, 78, 80
likelihood, 70
limitations, 18, 63, 74
linear, 5, 33, 44
litigation, 16, 36, 75, 77
living standard, 4, 8
living standards, 4
local government, 3, 17, 19, 23, 30, 31, 51
location, 47
long-term, 4, 6, 32, 35, 42

M

macroeconomic, 4, 31
magnetic, iv
maintenance, 8, 24, 25
management, 4
manufacturer, 14, 39, 58
manufacturing, 2, 8, 16, 19, 20, 21, 22, 23, 24, 25, 26, 30, 36, 40, 45, 46, 47, 49, 52, 53
manufacturing companies, 19
market, ix, 4, 6, 7, 9, 14, 16, 22, 35, 39, 44, 46, 55, 57, 58, 59, 62, 65, 67, 69, 71, 73
market share, 4
market value, 14, 39
marketing, 44, 57
markets, ix, 44, 46, 56, 59, 62, 63, 68
Massachusetts, 65
Massachusetts Institute of Technology, 65
matching funds, 43
MCC, 35
measures, 2, 6, 22, 47
military, 46
Minnesota, 65
missions, 73
MIT, 65, 81, 86
modernization, 43
money, 42, 58, 74

monopoly, 59, 69
monopoly power, 69
motivation, ix, 56, 60
movement, 41

N

nation, 4, 7, 8, 13, 17, 35, 44, 75
national, vii, 1, 4, 6, 7, 9, 35, 40, 57, 58, 61, 70, 71, 75
National Academy of Sciences, 67, 82
National Aeronautics and Space Administration, 17
National Defense Authorization Act, 18, 42
National Institute of Standards and Technology, vii, 2, 8, 11, 19, 24, 25, 30, 36, 41, 45, 52
National Institute of Standards and Technology (NIST), 11, 24, 25, 30
National Institutes of Health, 67, 87
National Science and Technology Council, 6, 8
National Science Foundation, 8, 35, 40, 49, 64, 75, 80, 81, 82, 84
negative consequences, 78
New York, iii, iv, 85
NIH, 67, 72, 74, 76, 77, 78, 80, 84, 87, 88
NIST, 3, 19, 21, 24, 25, 26, 30, 37, 42
non-profit, 3, 13, 23, 31, 49, 51, 52, 56
not-for-profit, 15, 40
nuclear, 63

O

obligations, vii, 1, 73, 74
Office of Science and Technology Policy, 87
offshore, 78
Omnibus Trade and Competitiveness Act, 11, 19, 36, 41
openness, ix, 56, 68
opposition, 43
organization, 3, 4, 13, 24, 25, 30, 35, 53, 62, 63

organizations, 13, 15, 19, 32, 37, 40, 62, 64, 69
outsourcing, 68
ownership, ix, 32, 47, 55, 56, 60, 62, 65, 69, 70, 72, 77, 79

P

paper, 57, 82, 83
partnership, 18, 42
partnerships, 21, 47, 73
Patent and Trademark Office, 78, 82
patents, 5, 16, 17, 18, 41, 57, 59, 63, 65, 67, 69, 70, 71, 72, 73, 74, 75, 77, 78, 79
Pennsylvania, 74
pension, 61
pension plans, 61
performance, 14, 22, 39, 68, 72
permit, viii, 29, 71
pharmaceutical, ix, 56, 57, 76, 77, 84
pharmaceuticals, 63, 68, 69, 71, 76, 79
planning, 6
play, 31, 61
policy levels, 6
portfolio, 57, 65
power, 35, 44, 69
preference, 22
President Bush, 8, 10, 12, 38, 39
President Clinton, 6, 8, 11
prices, 76
priorities, viii, 2, 34, 60, 74
private, vii, viii, ix, 1, 2, 4, 5, 6, 7, 8, 9, 10, 11, 13, 15, 16, 17, 18, 19, 22, 23, 29, 30, 31, 32, 35, 36, 40, 41, 42, 43, 46, 47, 49, 50, 52, 56, 57, 58, 60, 68, 69, 70, 72, 73, 76, 78
private sector, vii, viii, ix, 2, 4, 5, 6, 7, 8, 9, 10, 11, 13, 15, 16, 17, 18, 19, 22, 23, 29, 30, 31, 32, 35, 36, 40, 41, 42, 43, 46, 47, 49, 50, 52, 56, 57, 58, 60, 68, 69, 70, 72, 73, 78
private sector investment, 11, 30, 36, 50
private-sector, vii, 1
producers, 32
production, 16, 31, 36, 60, 70

Index

production costs, 31
productivity, vii, 1, 4, 5, 8, 33, 46, 70, 71, 79
profit, 3, 13, 15, 23, 31, 40, 49, 51, 52, 56, 74
profits, 34, 69, 71, 79
program, 2, 3, 5, 10, 11, 12, 13, 17, 18, 19, 20, 21, 23, 24, 25, 31, 35, 36, 37, 38, 39, 40, 41, 42, 45, 47, 49, 51, 52, 68
proliferation, 42
promote, viii, ix, 2, 4, 8, 13, 14, 15, 17, 23, 29, 35, 41, 42, 43, 49, 52, 56, 57, 58, 62, 64, 67, 68, 70, 74
promote innovation, 14
property, iv, ix, 2, 8, 17, 41, 47, 56, 57, 60, 65, 66, 69, 71, 74, 76, 77, 78
property owner, 60
proposition, 62
propulsion, 63
protection, 2, 8, 58, 59, 69, 74, 75, 77
public, ix, 7, 22, 41, 47, 56, 57, 58, 59, 62, 63, 65, 66, 68, 69, 71, 75, 76, 78, 79
public domain, 66, 78
public health, 58
public interest, 57, 71, 78, 79
public investment, 69
public policy, 22
public sector, ix, 56, 68, 79
Puerto Rico, 19

Q

quality improvement, 9

R

R&D, v, viii, ix, 3, 5, 6, 8, 9, 10, 13, 14, 15, 16, 17, 18, 22, 23, 24, 25, 27, 29, 30, 32, 33, 34, 35, 36, 40, 41, 42, 43, 44, 46, 47, 49, 52, 53, 55, 56, 57, 58, 60, 61, 62, 63, 64, 65, 66, 67, 68, 69, 70, 71, 72, 74, 75, 76, 79, 80, 82, 84, 87
range, 4, 7, 8, 40
rate of return, 5, 70
recession, 61
recognition, 18, 19, 33, 59
reduction, 2
Reform Act, 9, 14, 39
regional, 19, 30, 41, 50, 51, 61, 65
regulations, 9, 74
relationship, 7, 61
relationships, 34, 60, 74
Republican, 2, 22, 47
Republicans, 22
rescission, 3, 11, 12, 20, 21, 27, 37, 38
research, vii, viii, ix, 2, 3, 5, 8, 9, 10, 13, 14, 15, 17, 21, 23, 24, 25, 26, 27, 29, 30, 31, 32, 33, 34, 35, 39, 40, 41, 44, 46, 47, 48, 49, 50, 51, 52, 53, 54, 55, 56, 57, 58, 59, 60, 61, 62, 63, 64, 65, 66, 67, 68, 69, 70, 71, 72, 73, 74, 75, 76, 77, 78, 80
research and development, vii, viii, ix, 3, 5, 15, 16, 17, 24, 29, 30, 31, 32, 40, 41, 46, 51, 52, 55, 56, 57, 58, 63, 68, 69, 72, 75, 80, 87
Research and Development (R&D), 40
research funding, ix, 55, 59, 64, 76
researchers, 9, 15, 32, 41, 73, 75
resistance, 6
resolution, 70
resources, viii, 6, 7, 15, 16, 17, 18, 22, 30, 31, 33, 34, 35, 42, 45, 65, 74
responsibilities, vii, viii, 1, 6, 7, 16, 29, 32, 72
retention, 79
returns, 5, 6, 65, 70
rewards, 73, 74
risk, 8, 13, 58, 70
risks, viii, 6, 15, 29, 32, 34, 60, 62
royalties, 65, 73
royalty, ix, 17, 18, 56, 60, 71
royalty free, 71
rural, 21

S

safeguards, 79
safety, 63, 80
salaries, 61, 79
sales, 67
saving lives, 80
scarce resources, 22

Schiff, 50
school, 65, 66
science education, 3, 31
scientific community, 75
scientific progress, 73
scientists, 18, 34, 35, 60, 72, 73, 74, 75, 77, 78
security, 35, 40
seed, 11, 30, 36, 68
semiconductor, 40, 44, 46, 59
Senate, 2, 3, 10, 12, 13, 15, 20, 21, 23, 24, 25, 26, 27, 31, 38, 39, 48, 49, 50, 51, 52, 53, 54, 58, 67, 69, 81, 85, 88
separation, 14
series, 5, 12, 13, 21, 33, 39, 41, 57
services, iv, viii, ix, 2, 3, 4, 5, 8, 14, 22, 23, 29, 31, 32, 34, 44, 45, 47, 51, 56, 60, 61, 70
shape, 34, 60, 72
sharing, viii, ix, 13, 18, 29, 56, 68
short-term, 42
signals, 42
skills, viii, 15, 30, 33, 35, 41
Small Business Administration, 61
small firms, 10, 37, 58, 61, 62, 63, 67, 68
social costs, 59
software, 36
solutions, 70
speed, 33, 47
spin, 71, 73
SPR, 43
stages, 44
standard of living, 4, 31
standards, 70
State of the Union, 3, 8, 31
statistics, 54, 82, 84, 87, 88
statutory, 20
steel, 40
stimulus, 70
strategic, 7, 68
strategies, 2, 8
strength, 4, 59
subsidization, 69
superiority, 7
suppliers, 33
syndrome, 17
systems, 46, 47, 72

T

talent, 73, 75
tax credit, 3, 5, 8, 9, 14, 24, 26, 27, 31, 39, 48, 49, 50, 51, 52, 53, 54, 68
tax credits, 5
tax deduction, 14, 39
tax incentive, 8, 50, 57, 72
tax incentives, 8, 50, 57, 72
tax preferences, 22
taxes, 71, 79
taxpayers, 69
teachers, 8
technical assistance, 30, 42
technicians, 18
technological advancement, vii, viii, 1, 2, 5, 6, 7, 9, 23, 29, 31, 33, 34, 35, 42, 57, 60, 61, 70
technological change, 61
technological progress, viii, 2, 4, 14, 31, 70
technology, vii, viii, ix, 1, 2, 3, 4, 5, 6, 7, 8, 9, 10, 14, 15, 16, 17, 18, 19, 21, 22, 26, 29, 30, 31, 33, 35, 36, 40, 41, 42, 43, 44, 46, 47, 53, 56, 57, 58, 60, 61, 63, 64, 65, 66, 67, 68, 69, 71, 72, 74, 77, 78, 79
Technology Administration, 6, 20, 87
technology commercialization, 58
technology transfer, ix, 5, 17, 18, 19, 30, 41, 46, 56, 66, 68, 72, 74, 79
tenure, 74
testimony, 67, 77
Texas, 76
The Economist, 59, 65, 81
theory, 5
therapeutics, 71, 76
thinking, 23
threat, 73
time, 2, 11, 15, 18, 40, 43, 58, 59, 61, 62, 63, 65, 69, 77
time constraints, 19
time consuming, 59
time frame, 15, 40, 62, 63
TIP, 13, 24, 25
title, viii, ix, 15, 17, 18, 30, 40, 41, 56, 58, 60, 61, 62, 63, 64, 70, 79

Index

tracking, 72
trade, 4, 31, 46
training, 14, 39, 60
transaction costs, 77
transcript, 82
transfer, viii, ix, 3, 5, 9, 16, 17, 18, 19, 23, 30, 31, 41, 42, 44, 46, 51, 56, 57, 66, 68, 72, 74, 79
transition, 65
trend, 9
trusts, 75

U

U.S. economy, 4, 7, 47, 61
uncertainty, 6
uniform, 58
United States, vii, viii, 4, 5, 7, 14, 15, 16, 29, 36, 39, 41, 44, 47, 57, 60, 62, 63, 66, 71, 80, 84, 85
universities, viii, 3, 9, 10, 11, 13, 14, 15, 17, 18, 23, 29, 30, 31, 32, 34, 35, 36, 39, 40, 41, 49, 51, 58, 60, 62, 63, 64, 65, 66, 67, 68, 70, 74, 75, 76, 77, 78, 79, 80
unobligated balances, 21

Utah, 65

V

venture capital, 66, 68, 75
Vermont, 80

W

wages, 9, 79
Wall Street Journal, 86
war, 34, 60
weapons, 63
web, 82, 86, 87
welfare, 35, 79
well-being, 79
Wisconsin, 79
World War, 14, 34, 60
World War I, 14, 34, 60
World War II, 14, 34, 60
WP, 82
writing, 61